Asterisms
for Small Telescopes and Binoculars

John C. Raymond

BRUNSWICK PUBLISHING CORP
LAWRENCEVILLE, VIRGINIA

Copyright © 2005 by John C. Raymond

All rights reserved under International and Pan-American Copyright Conventions. No part of this book may be reproduced in any form or by any means, electronic or mechanical, including photocopying or by any informational storage or retrieval systems, without written permission from the author and the publisher, except by a reviewer who may quote brief passages in a review.

SKYTOOLS™ 2 data and charts used with permission from CappellaSoft, www.Skyhound.com

Library of Congress Cataloging-in-Publication Data

Raymond, John C., 1970-
   Asterisms for Small Telescopes and Binoculars / John C. Raymond
   p. cm.
   Includes bibliographical references.
   ISBN-13: 978-1-55618-205-1 (pbk. : acid-free paper)
   ISBN-10: 1-55618-205-8 (pbk. : acid-free paper)
     1. Stars—Clusters—Charts, diagrams, etc.   2. Constellations—Charts, diagrams, etc.   I.Title
QB851.R39 2005
523.8'022'3—dc22
                                                                        2005053100

Manufactured in the United States of America
First edition
Published by Brunswick Publishing Corp., PO Box 555, Lawrenceville, VA 23868, www.brunswickbooks.com

Acknowledgments

This book was inspired by:

Finder Charts of Overlooked Objects by Brent Watson
Published by Sky Spot
1263 East Beverly Way, Bountiful, UT, www.sky-spot.com

Burnham's Celestial Handbook, Volumes 1, 2, & 3 by Robert Burnham Jr.
Dover Publications
31 East 2nd Street, Mineola, NY

The author wishes to thank his family for their love and understanding.

Also: Friends and fellow observers of the Richmond Astronomical Society, Back Bay Amateur Astronomers, Rappahannock Astronomy Club, East Coast Star Party, Mid Atlantic Star Party, Mason Dixon Star Party, and Big Meadows Star Party. Phil Harrington for all his wonderful books.

Special thanks to Scott Anderson, Kent Blackwell, Mark Ost, and George Reynolds for reviewing a rough draft of this book. Their comments and criticisms have been very helpful.

The author also found the following magazines helpful: *Amateur Astronomy, Astronomy, Sky and Telescope*, and *Odyssey*.

Table of Contents

Introduction 1	Struve 1835 Finder 35	Canis Minor
Types of Asterisms 2		Delta 2 Detail 70
The Charts 3	Camelopardalis	Delta 2 Finder 71
Bibliography 4	Cam. Ast. Detail 36	
Observing Asterisms 5	Cam Ast. Find 37	Cancer
		Cascade Detail 72
Andromeda	Kemble's Cascade:	Cascade Finder 73
8 and 11 detail 6	Detail to 9th Mag. 39	Sigma Detail 74
8 and 11 finder 7	Detail to 7th Mag. 40	Sigma Finder 75
Omicron Detail 8	Finder 41	
Omicron Finder 9		Coma Bereneces
	Cassiopeia	M53 Ast, Detail 76
Aquarius	1 – 2 Cas Detail 42	M53 Ast. Finder 77
Waters Detail 10	1 – 2 Cas Finder 43	
Waters Detail Cont' 11	52 – 53 Cas Detail 44	Canes Venatici
Delta finder 12	52 – 53 Cas Finder 45	AW CVn Detail 78
Tau finder 13	Rho Cas Detail 46	AW CVn Finder 79
98 – 101 finder 14	Rho Cas Finder 47	
88 – 86 finder 15	Sigma Cas Detail 48	Corvus
Omega finder 16	Sigma Cas Finder 49	Corvus and Draco 38
103 – 108 finder 17	Theta Cas Detail 50	Struve 1659 Detail 80
Waters Detail Cont' 18	Theta Cas Finder 51	Struve 1659 Finder 81
PsiAqr finder 19	Upsilon Cas Detail 52	
M73 detail 20	Upsilon Cas Find 53	Corona Borealis
M73 finder 21	Zeta Cas Detail 54	Tau CrB Detail 82
Water Jar detail 22	Zeta Cas Finder 55	Tau CrB Finder 83
Water Jar finder 23		
	Cepheus	Cygnus
Aries	15 Cep Detail 56	31 Cyg Detail 84
Xi Ari detail 24	15 Cep Finder 57	31 Cyg Find 85
Xi Ari finder 25		
	Cetus	Draco
Auriga	Eta Cet Detail 58	Cross Detail 86
Island Detail 26	Eta Cet Finder 59	Cross Finder 87
Island Finder 27	49 – 50 Cet Detail 60	Kappa Dra Detail 88
Kids Detail 28	49 – 50 Cet Finder 61	Kappa Dra Finder 89
Kids Finder 29	Zeta Cet Detail 62	Lambda Dra Detail 90
Reverse Kids Detail 30	Zeta Cet Finder 63	Lambda Dra Find 91
Rev. Kids Finder 31	Upsilon Cet Detail 64	Pseudocluster Det 92
	Upsilon Cet Finder 65	Pseudocluster Fnd 93
Boötes	Xi – 64 Cet Detail 66	Kemble 2 detail 94
Kappa – Iota Detail 32	Xi – 64 Cet Finder 67	Kemble 2 finder 95
Kappa – Iota Find 33	77 – 80 Cet Detail 68	UX Dra Detail 96
Struve 1835 Detail 34	77 – 80 Cet Finder 69	UX Dra Finder 97

Table of Contents, cont'd

Eridanus
5 – 7 Eri Detail 98
5 – 7 Eri Finder 99
Rho Eri Detail 100
Rho Eri Finder 101

Fornax
Chi For Detail 102
Chi For Finder 103

Hercules
Chi Her Detail 104
Chi Her Finder 105
60 Her Detail 106
60 Her Finder 107

Hydra
1 – C – 2 Hya Detail 108
1 – C – 2 Hya Find 109
I Hya Detail 110
I Hya Find 111

Lacerta
Little Cas Detail 112
Little Cas Find 113

Leo
Tau Leo Detail 114
Tau Leo Finder 115

Leo Minor
22 LMi Detail 116
22 LMi Finder 117
30 LMi Detail 118
30 LMi Finder 119
46 LMi Detail 120
46 LMi Finder 121

Libra
Libra Hydra Detail 122
Libra Hydra Find 123
Pseudocluster Det. 124
Pseudocluster Fnd 125

Lupus
Theta Lup Detail 126
Theta Lup Finder 127

Ophiuchus
Yed Prior Detail 128
Yed Prior Finder 129
Kappa Oph Detail 130
Kappa Oph Finder 131

Pegasus
Enif Detail 132
Enif Finder 133
35 Peg Detail 134
35 Peg Finder 135
Kappa Peg Detail 136
Kappa Peg Finder 137
12 Peg Detail 138
12 Peg Finder 139

Pisces
Little Water Jar Detail 140
Little Water Jar Finder 141
Y Detail 142
Y Finder 143

Scorpius
Omega Sco Detail 144
Omega Sco Finder 145

Sagitta
13 Sge Detail 146
13 Sge Finder 147
15 Sge Detail 148
15 Sge Finder 149

Sagittarius
Jewels Detail 150
Jewels Finder 151
M22 Ast. Detail 152
M22 Ast. Finder 153

Ursa Major
82 UMa Detail 154
82 UMa Finder 155

Ursa Minor
Delta UMi Detail 156
Delta UMi Finder 157
Eta Umi Detail 158
Eta Umi Finder 159

Umi PSCL A
Detail 160
Finder 161

Umi PSCL B
Detail 162
Finder 163

Virgo
Coronae Vir Detail 164
Coronae Vir Finder 165
65 – 66 Vir Detail 166
65 – 66 Vir Finder 167

M104 Eastern
Detail 168
Finder 169

M104 Western
Detail 170
Finder 171
Syrma Detail 172
Syrma Finder 173

Vulpecula
Coathanger Detail 174
Coathanger Finder 175

Introduction

Why asterisms?

I can't explain why I'm fascinated by asterisms, any more than I can explain my passion for astronomy. It all started in August of 2002. On Aug 6th, I was observing one of my favorite doubles, 32 Camelopardalis (Struve 1694). I decided to peruse the surrounding area — just moving the scope around in a random search of an area I had never observed before. I found what I described in my notes as an "unmarked cluster." Research revealed no mention of the Camelopardalis Asterism in any book or on the Internet — and it's visible to the eye in a dark location! Later in October, I spotted the Enif asterism while looking for M15. This is a purely telescopic asterism of 10th and 11th magnitude stars. It's located just north of Enif — an important star when finding M15.

In April of 2003, I was observing under the dark skies of Powhatan, Virginia. I noticed something. Some stars appeared "fuzzy." Kappa Draconis is a good example. The proximity of 4 and 6 Draconis, 6th mag. stars just at the threshold of visibility, give this group of stars a visual fuzzy appearance. I decided then to investigate these visual fuzzy objects.

Visual fuzzy objects are not new, the most famous being the bright Messier objects: M44 in Cancer, M31 in Andromeda, M42 in Orion, M45 the Pleiades, M8 in Sagittarius, M7 in Scorpius. The Double cluster in Perseus is also a conspicuous fuzzy object. The Milky Way is full of clusters and nebulae visible to the naked eye.

My observing program changed from the standard recipe of double stars and deep sky objects listed in star charts to every visual fuzzy object visible to my eye, as well as asterisms seen in the finder scope or eyepiece. Now I search for asterisms in every constellation. I found hundreds, but many are faint, sparse, or require some imagination to see. Many are simple pairs of bright stars or pairs where one has a noticeable color. Some are polygonal shapes, lines or curves. Very few asterisms resemble ordinary objects.

The purpose of this book is to share my enthusiasm for astronomy with my fellow observers. I want to show that the sky has much more to offer than traditional observing dictates. I also encourage other astronomers to explore the night sky and find something new every time.

My observing philosophy can be found on page 54 of *Burnham's Celestial Handbook*.

JOHN C. RAYMOND
Richmond, Virginia
June 2005

Types of Asterisms

There are several types of asterisms. Many asterisms fit into more than one category. Asterisms are not scientific or official, but generally agreed upon by multiple observers.

1. Grand Asterisms: Formed by the brightest stars of the season from different constellations. Examples:
   Winter Triangle, Arc to Arcturus, Summer Triangle, Great Square (of Pegasus and Andromeda), Winter Hexagon

2. Classical Asterisms: Part of one constellation, often listed in star charts and observing books. Ex.: Big Dipper, Sickle of Leo, Teapot of Sagittarius, and Orion's Belt.

3. Consistent Asterisms: A subset of stars within a constellation that are a logical part of that constellation. Ex.: Orion's Belt, Orion's Sword, Head of Draco, Bow and Arrow of Sagittarius, Kids in Auriga. Human figures are expected to have clothing and accouterments. Living things are expected to have body parts, wings, horns, or tails.

4. Inconsistent Asterisms: A subset of stars within a constellation that have no logical relationship with the figure depicted by the constellation. Ex.: Keystone of Hercules, Big Dipper, Sickle of Leo, Teapot of Sagittarius. A keystone is not a logical part of the human figure of Hercules. A dipper has no relationship to the bear, Ursa Major.

5. Obsolete Asterisms: Formed of no longer recognized constellations. Ex.: Poiniatowski's Bull* (in Ophiuchus).

6. Binocular Asterisms: Best seen in Binoculars. Ex.: Engagement Ring of Polaris*, Coathanger (Collinder 399), Kemble's Cascade.

7. Telescopic Asterisms: Small, faint groups of stars that resemble a constellation or everyday object, or asterisms formerly cataloged as clusters. Listings of the NGC open clusters include many of these. Also includes the temporary asterisms used by deep sky observers to find faint objects.

*Not described in this book.

The Charts

I made the charts using SKYTOOLS™ 2 software. I have attempted to show the asterisms in detail and name the stars composing it. I dislike anonymous stars! The stars are like people — they all have names. Even if that name is a catalog number. I have included in this work the best and brightest asterisms I have observed.

Each chart has two parts: a finder chart for locating the asterism and a close-up, detail chart.

The finder chart lists the name of the asterism, constellation, brightest star, and coordinates of the brightest star. The finder chart simulates the naked eye, finderscope, and eyepiece views using various small aperture instruments. The eyepiece view of the finder charts has a dashed circle to represent the eyepiece field of view, with an arrow indicating west. The eyepiece view orientation will vary depending on the instrument used.

The finder charts are based on four different optical instruments: 7 x 50 binocular, 16 x 70 binocular, 80 mm refractor, 6 inch reflector. These are instruments I have used for observing, but are not always the ones I saw the asterism with.

All the finderscope views simulate the popular Telrad™ finder. One of the great features of SKYTOOLS™ is the ability to make finder charts based on a variety of finders.

The detail chart is a close up view of the chart. The asterism is described briefly, and data are given for the stars that comprise the asterism. The data are intended to answer questions at the eyepiece: How far? Is it a double? What color is it? The brightest star of the asterism is denoted with an asterisk. This star's coordinates are given at the top of the page.

What is seen in the eypiece will vary from what is printed on the charts. Many factors are involved , including aperture, sky transparency, light pollution, and the observer's visual acuity.

SKYTOOLS™ 2
Observation Planning, Charting, and Logging
www.skyhound.com

Bibliography

Hirshfield, A., R.W. Sinnott, and F. Ochsenbein, editors, *Sky Catalog 2000.0 Vol. 1, Stars to Magnitude 8.0, 2nd edition*, Sky Publishing Corp., Cambridge, MA.

Hirshfield, A., and R.W. Sinnott, editors, *Sky Catalog 2000.0 Vol. 2, Double Stars, Variable Stars, and Nonstellar Objects,* Sky Publishing Corp., Cambridge, MA, 1985.

Kepple, G., Sanner, G., *The Night Sky Observer's Guide Vol. 1, Autumn and Winter*, Wilmann-Bell, Richmond, VA, 1998.

Kepple, G., Sanner, G., *The Night Sky Observer's Guide Vol. 2, Spring and Summer*, Wilmann-Bell, Richmond, VA, 1998.

Ridpath, I., and W. Tirion, *Stars and Planets*, third edition, Princeton University Press, Princeton, NJ, 2001

Tirion, W., *Sky Atlas 2000.0*, Sky Publishing Corp, Cambridge, MA, 1981.

Tirion, W., B. Rappaport, and W. Remaklus, *Uranometria 2000 Vol. 1, The Northern Hemisphere to -6°,* Willmann-Bell, Richmond, VA, 2001

Tirion, W., B. Rappaport, and W. Remaklus, *Uranometria 2000 Vol. 2, The Southern Hemisphere to +6°,* Willmann-Bell, Richmond, VA, 2001

Observing Asterisms

Asterisms are easy to find and observe. Who has not seen the Big Dipper or Orion's Belt? The familiar constellation figures themselves are asterisms, unrelated stars appearing close together in the sky by our perspective.

The Authorities in the astro community shake their heads in disapproval at the mention of asterisms. "Those stars are not gravitationally bound! They are not related!" High school physics tells us that all matter is gravitationally bound. All the visible stars are part of the Milky Way. The are related to you and me, who are also part of our galaxy.

Observing asterisms requires imagination. Many observers say the faint galaxy observers have good imaginations, for seeing the faintest of deep sky objects. Asterism observing is the opposite. The observer is required to see the sky background itself as an "object" worthy of observation. I cannot observe M104 without peeking at the asterisms that flank it. Traditional observing celebrates the "object" and relegates the sky background to obscurity.

One great attribute of asterisms is that many can be appreciated with various apertures and instruments. The Kappa Draconis Asterism is a good example. Casual eye observing sees this star between the Big Dipper and Polaris. Careful visual scrutiny sees a fuzzy star. Low power, small aperture instruments reveal the proximity of other bright stars. A large aperture scope shows the star's contrasting colors in their glory. Many observers will disagree, stating that "they are just stars," not "spectacular" deep sly objects. That's fine, the sky is big enough for everyone to find something exciting to observe.

Asterism test: Center the suspected asterism in the eyepiece. Move the telescope until the object is out of the field, and then bring it back. Repeat this but at right angles to the original movement. If I still see something there, if the stars appear to belong together, I log it. I generally avoid looking for asterisms that are near the galactic equator, as the stars there are so concentrated. I have made some exceptions, as you will see.

Name	**The 8 & 11 Andromedae Asterism**
Location	RA: 23h 17m 44s, Dec: +49° 00' 55" (2000) in Andromeda
Colorful Stars	8, 11 And, HIP 115171, 115363
Found With	Direct Vision
Description	To the eye 8 and 11 And appear as a fuzzy spot in the Milky Way in western Andromeda near Lacerta. Closer inspection shows four fainter stars in the field. The distance between 8 and 11 is 29 arc minutes, almost half a degree.

[Star chart showing 8 And A, HIP 115128 A (72), HIP 115363 (74), 11 And (54), HIP 115114 A (76), HIP 115171 A (63)]

Principal Stars

*8 And Multiple Star System
aka HR 8860, HD 219734, SAO 52871,
BU 717, ADS 16656
Mag: V 4.76 to V 4.93, M2III
Distance: 650 +/- 85.0 ly
AB: 4.82+13 mag
In 1878: PA 161° Sep 7.60"
AC: 4.82+10.3 mag,
In 1880: PA 131° Sep 219.40"
AD: 4.82+11.6 mag,
In 1916: PA 232° Sep 54.90"

HIP 115128 Multiple Star System
aka HD 219917, SAO 52899, PPM 64064
Magnitude: 7.24, A2
Distance: 450 +/- 77.0 ly
AB: 7.24+9.2 mag, B=SAO 52895;
In 1881: PA 231° Sep 49.40"
AC: 7.24+11.2 mag;
In 1991: PA 107° Sep 0.54"
Distance: 330 +/- 21.0 ly

11 And Single Star
aka HR 8874, HD 219945, SAO 52907
Magnitude: 5.44, K0

HIP 115114 Multiple Star System
aka HD 219890, BD +47 04107,
ADS 16673, STT 493, ADS 16673
Magnitude: 7.64
Distance: 1000 +/- 350.0 ly
AB: 7.64+11.33 mag, B=HIP 115114
In 1991: PA 25° Sep 8.34"

HIP 115171 Multiple Star System
aka HR 8875, HD 219962, SAO 52912,
PPM 64080, STT4244
Magnitude: 6.29, K0
Distance: 480 +/- 48.0 ly
AB: 6.29+9.8 mag, B=PPM 64076
In 1920: PA 300° Sep 85.80"
AC: 6.29+11.4 mag,
In 1925: PA 105° Sep 134.30"

HIP 115363 Single Star
aka HD 220274, SAO 52942, PPM 64124
Mag 7.4 K0
Distance 1100 +/- 340ly

Object data from SKYTOOLS™ 2 — used with permission.

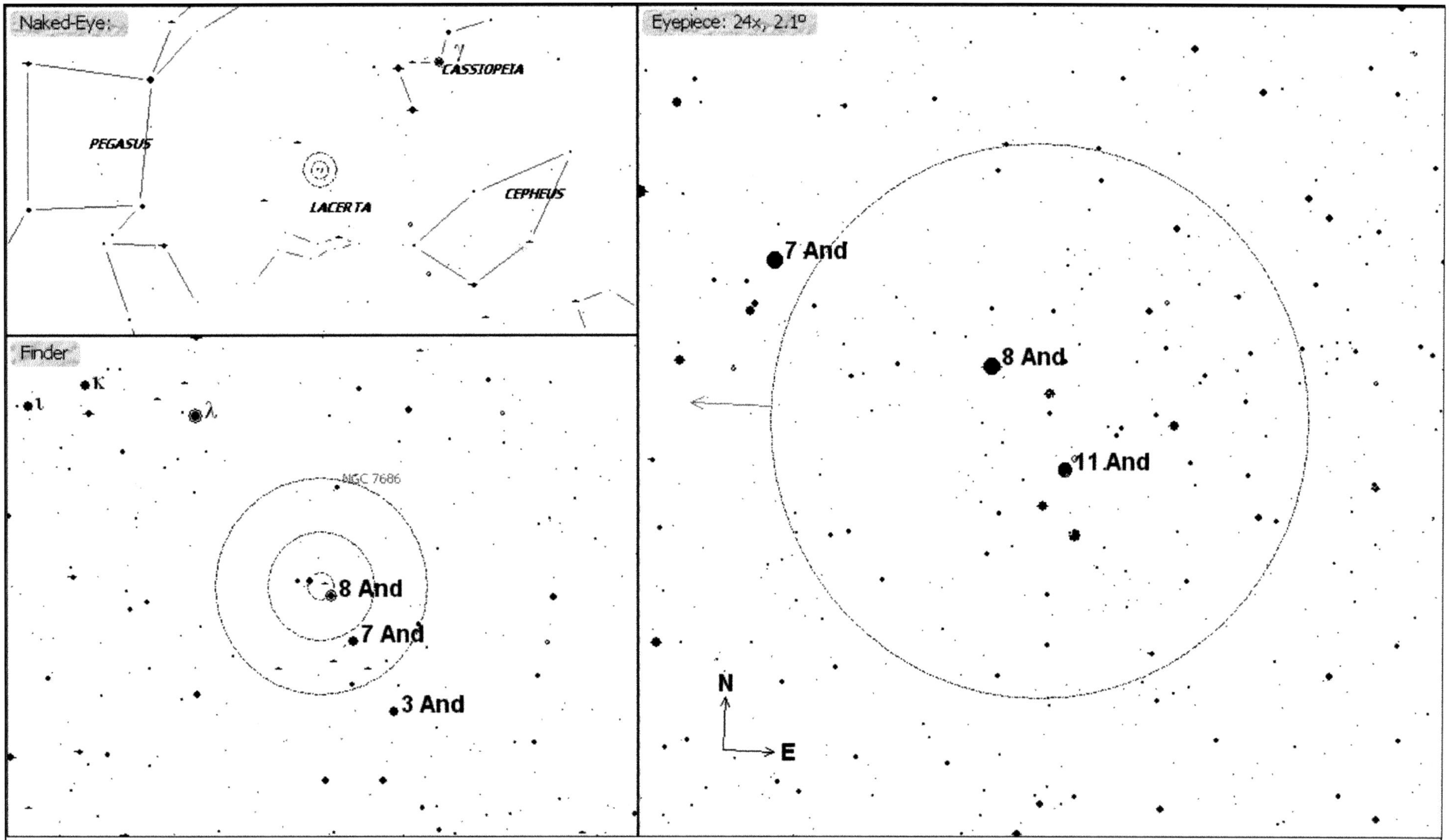

The 8 & 11 Andromedae Asterism / Andromeda / 8 And / RA: 23h 17m 44s, Dec: +49° 00' 55" (2000)

80mm f/6 Refractor

Name	**The Omicron (o) Andromedae Asterism**
Location	RA: 23h 01m 55s, Dec: +42° 19' 34" (2000) in Andromeda
Found With	Direct Observation
Description	To the eye, Omicron and 2 And appear as a fuzzy spot in the Andromeda Milky Way near Lacerta. The distance between the two stars is 27 arc minutes.

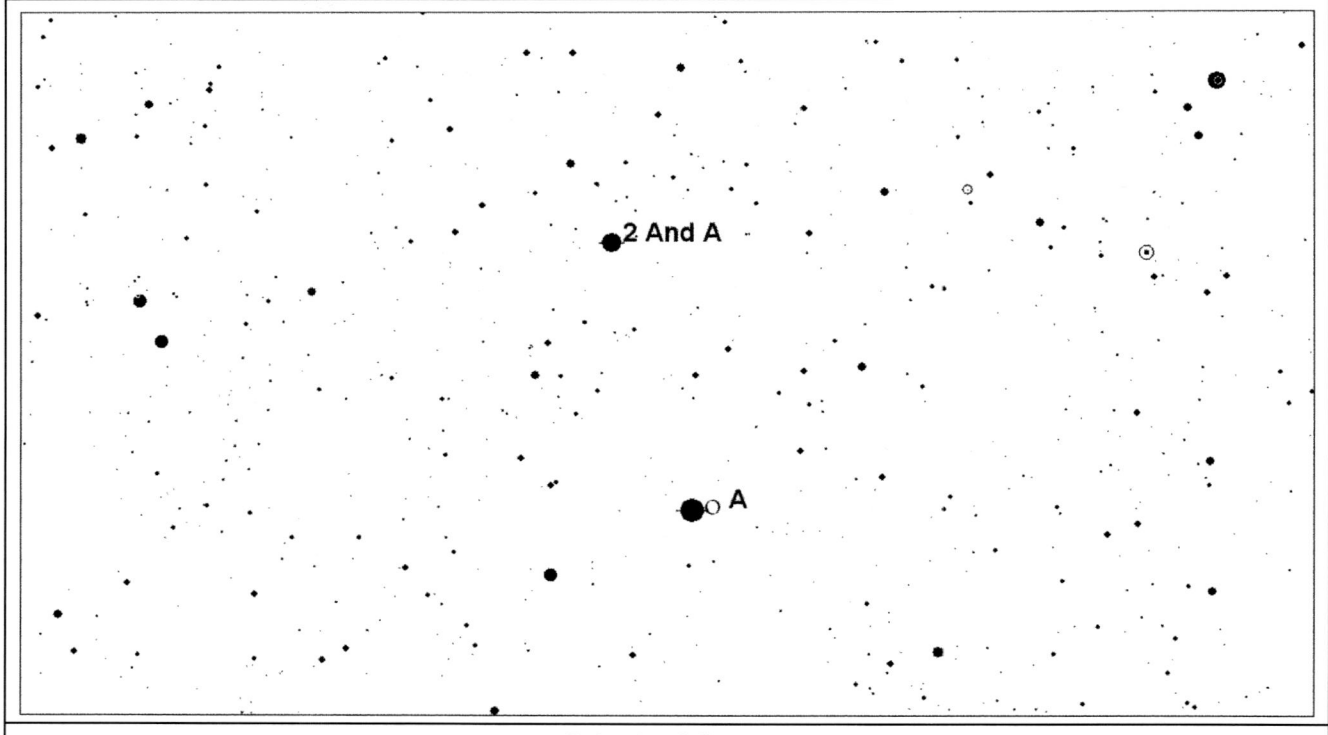

Principal Stars

*Omicron And Multiple Star System	**2 And** Multiple Star System
aka 1 And, Omicron And, GCVS 269, HR 8762, HD 217675, SAO 52609, PPM 63726, HIP 113726	aka HR 8766, HD 217782, SAO 52623, PPM 63742, BU 1147, ADS 16467
Variability: Type: GCAS	Magnitude: 5.09, F2
Mag: V 3.58 to V 3.78, B6III	Distance: 350 +/- 26.0 ly
Distance: 690 +/- 98.0 ly	AB: 5.09+7.43 mag, B=SAO 52623
AB: 3.62+6.03 mag;	Preliminary Orbit: P=76.6 yr, a=0.28"
In 1991: PA 345° Sep 0.20"	PA 1° Sep 0.37" (2004.8)
	AC: 5.09+13.7 mag
	In 1911: PA 192° Sep 90.40"
	AD: 5.09+12 mag,
	In 1985: PA 283° Sep 142.8"

Object data from SKYTOOLS™ 2 — used with permission.

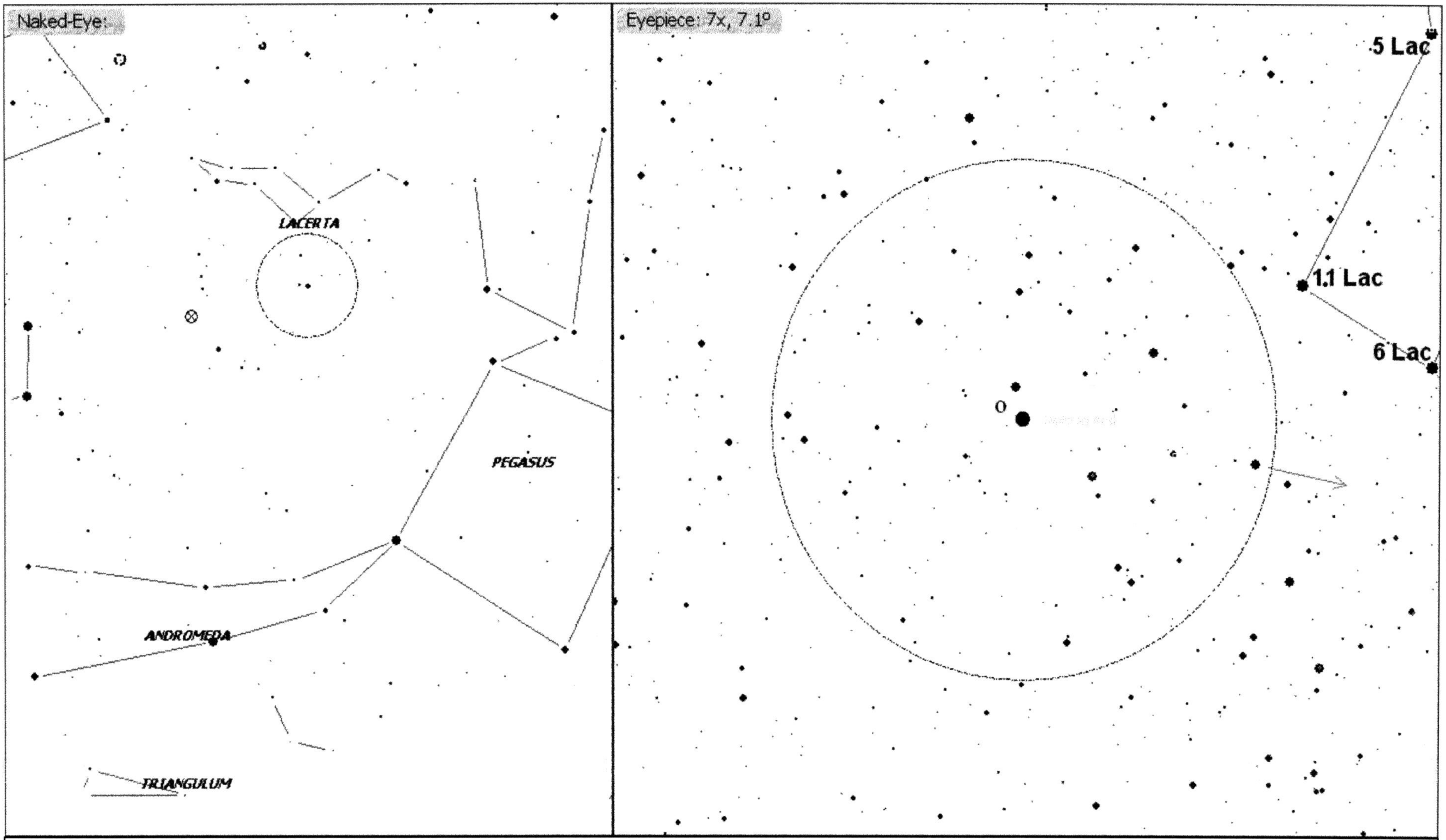

Omicron Andromedae Asterism / Andromeda / RA: 23h 01m 55s, Dec: +42° 19' 34"

7 x 50 Binocular

Name	**The Waters of Aquarius**
Location	22 to 24hr RA, -8° to -23° Dec
Found With	Naked Eye
Description	The Waters of Aquarius are the collection of asterisms that appear like water flowing downhill. The water is composed of seven groups of pairs or chains of stars. The asterisms are, clockwise: τ^1-τ^2, δ-77, 88-86, 98-101, 103-108, ω^1-ω^2, and $\psi^1\psi^2\psi^3$ Aquarii. North is up.

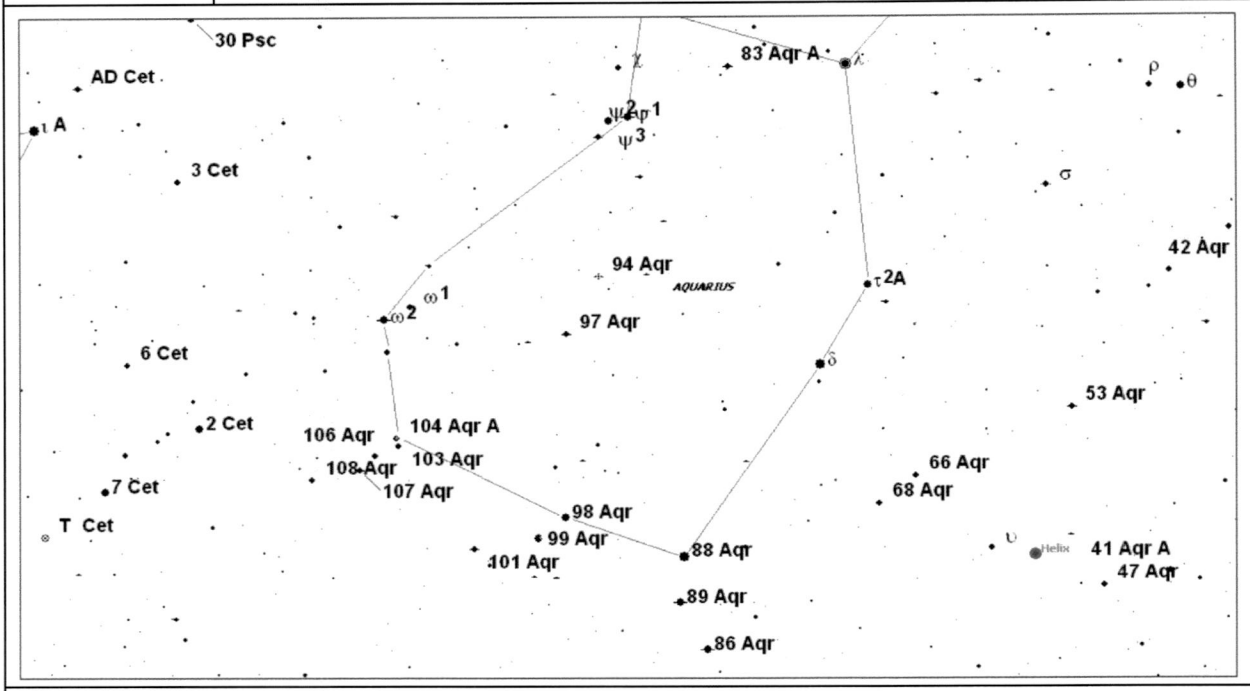

Principal Stars

Tau 2 Aqr Multiple Star System
aka 71 Aqr, HR 8679, HD 216032, SAO 165321
Magnitude: 3.98, K5III
Distance: 380 +/- 43.0 ly
AB: 4.05+8.7 mag,
In 1825: PA 293° Sep 133.40"

Tau 1 Aqr Multiple Star System
aka 69 Aqr, HR 8673, HD 215766, SAO 165298,
Struve 2943, ADS 16268
Magnitude: 5.68, B9
Distance: 260 +/- 25.0 ly
AB: 5.68+9 mag; In 1959: PA 121° Sep 23.70"

Delta Aqr Single Star
aka Skat, 76 Aqr, HR 8709, HD 216627,
SAO 165375
Magnitude: 3.28, A3
Distance: 160 +/- 18.0 ly

77 Aqr Single Star
aka HR 8711, HD 216640, SAO 165376
Magnitude: 5.53, K1III, Distance: 140 +/- 4.8 ly

88 Aqr Single Star
aka HR 8812, HD 218594, SAO 191683
Magnitude: 3.64, K1III
Distance: 230 +/- 16.0 ly

89 Aqr Multiple Star System
aka HR 8817, HD 218640, SAO 191687
Magnitude: 4.68, A3IV
Distance: 520 +/- 88.0 ly
AB: 4.71+6.04 mag;
In 1991: PA 285° Sep 0.20"

86 Aqr Multiple Star System
aka HR 8789, HD 218240, SAO 191651,
B 588, ADS 16511
Magnitude: 4.48, G8III
Distance: 190 +/- 9.6 ly
AB: 4.48+14.8 mag In 1925: PA 83° Sep 2.90"
AC: 4.48+6.8 mag; In 1991: PA 119° Sep 0.25"

98 Aqr Single Star
aka HR 8892, HD 220321, SAO 191858
Magnitude: 3.98, K0III
Distance: 160 +/- 5.8 ly

99 Aqr Variable Star
aka HR 8906, HD 220704, SAO 191900
Magnitude: V 4.35 to V 4.45, K4III
Distance: 310 +/- 21.0 ly

Object data from SKYTOOLS™ 2 — used with permission.

Waters of Aquarius

Principal Stars Continued

100 Aqr   Single Star
aka HR 8932, HD 221357, SAO 191970
Magnitude: 6.24, F0
Distance: 260 +/- 18.0 ly

101 Aqr   Multiple Star System
aka HR 8939, HD 221565, SAO 191988
Magnitude: 4.72, A0
Distance: 320 +/- 24.0 ly
AB: 4.7+7.43 mag;
In 1991: PA 125° Sep 0.84"

103 Aqr   Single Star
aka HR 8980, HD 222547, SAO 165834
Magnitude: 5.33, K4III
Distance: 580 +/- 79.0 ly

104 Aqr   Multiple Star System
aka HR 8982, HD 222574, SAO 165836, HJ 5413
Magnitude: 4.83, G2I
Distance: 640 +/- 110.0 ly
AB: 4.82+7.9 mag, B=HD 222561;
In 1919: PA 6° Sep 120.10"
BC: 7.9+11.9 mag
In 1913: PA 76° Sep 90.20"

106 Aqr   Single Star
aka HR 8998, HD 222847, SAO 165854
Magnitude: 5.24, B9
Distance: 330 +/- 29.0 ly

107 Aqr   Multiple Star System
aka HR 9002, HD 223024, SAO 165867, H 24, ADS 16979
Magnitude: 5.28, F0
Distance: 210 +/- 16.0 ly
AB: 5.28+6.72 mag, B=HIP 117218
In 1991: PA 136° Sep 6.79"

108 Aqr   Variable Star
aka ET Aqr, GCVS 783, HR 9031, HD 223640, SAO 165918
Type: ACV, Period: 3.730000 days
Magnitude: V 5.16 to V 5.21, A
Distance: 320 +/- 33.0 ly

Omega 2 Aqr   Multiple Star System
aka 105 Aqr, HR 8988, HD 222661, SAO 165842, BU 279, ADS 16944
Magnitude: 4.51, B9
Distance: 150 +/- 6.3 ly
AB: 4.49+10.6 mag
In 1875: PA 86° Sep 5.70"

Omega 1 Aqr   Single Star
aka 102 Aqr, HR 8968, HD 222345, SAO 165818
Magnitude: 4.97, A7IV
Distance: 130 +/- 4.7 ly

Psi 3 Aqr   Multiple Star System
aka 95 Aqr, HR 8865, HD 219832, SAO 146635, HO 199, ADS 16671
Magnitude: 4.99, A0
Distance: 250 +/- 19.0 ly
AB: 4.99+11.2 mag,
In 1959: PA 174° Sep 1.50"
AC: 4.99+13.3 mag,
In 1912: PA 230° Sep 130.10"

Psi 2 Aqr   Single Star
aka 93 Aqr, HR 8858, HD 219688, SAO 146620
Magnitude: 4.40, G8III
Distance: 320 +/- 33.0 ly

Psi 1 Aqr   Multiple Star System
aka 91 Aqr, HR 8841, HD 219449, SAO 146598, Struve 5012, ADS 16633, BU 1220
Magnitude: 4.25, K0III
Distance: 150 +/- 6.0 ly

Psi 1 Aqr

Component	Magnitudes	Date	Position Angle	Separation
AB	4.24+10.3	1836	312°	49.60"
BC	10.3+11.5	1958	105°	0.30"
AD	4.24+13.5	1924	274°	80.40"
BE	10.3+14.3	1924	341°	19.70"

Object data from SKYTOOLS™ 2 — used with permission.

The Delta (δ) Aquarii Asterism / Aquarius / RA: 22h 54m 39.0s, Dec: -15° 49' 15"

Finder

Eyepiece: 24x, 2.1°

80mm f/6 Refractor

The Tau (τ) Aquarii Asterism / Aquarius / RA: 22h 49m 36s, Dec: -13° 35' 33"

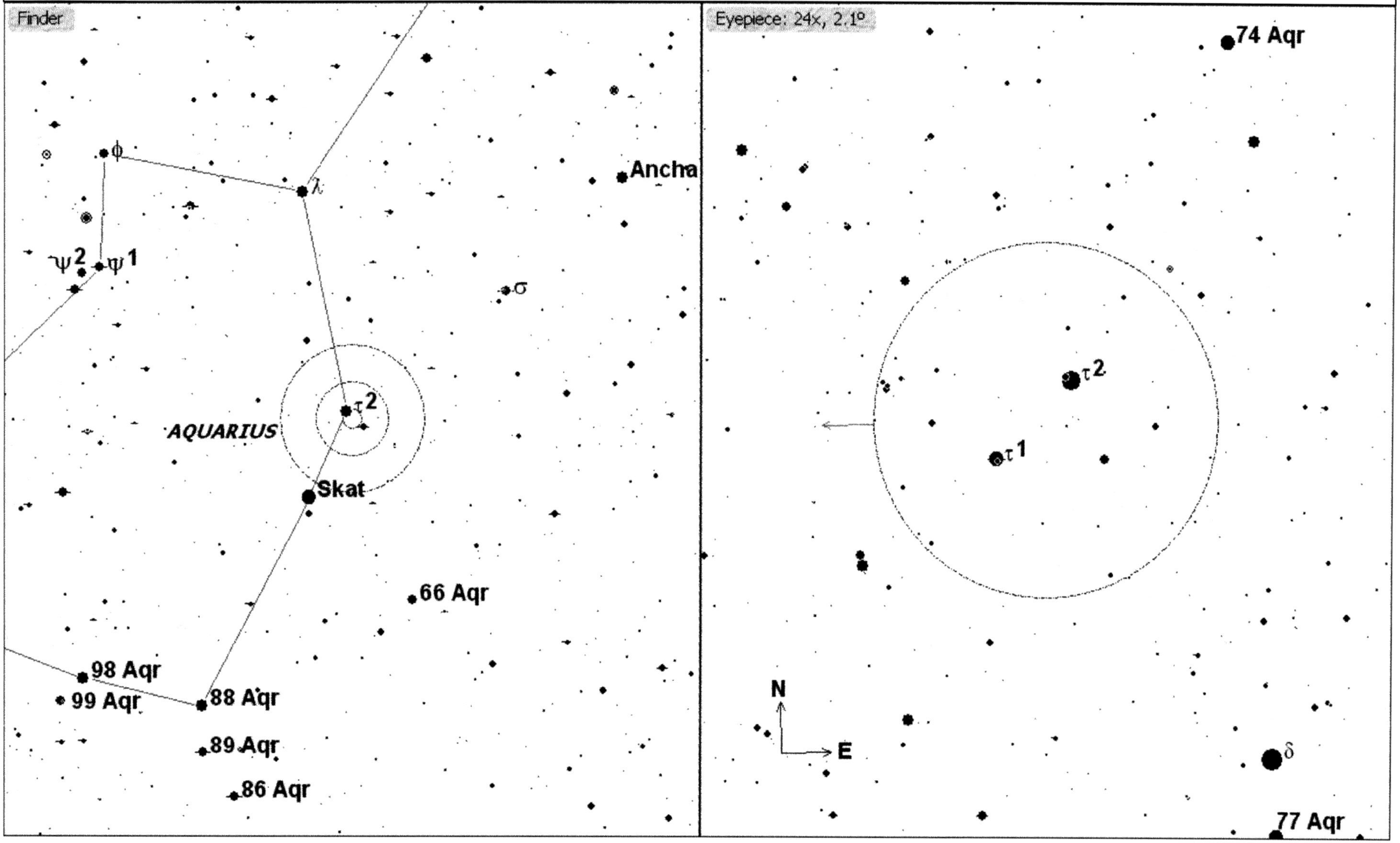

Finder

Eyepiece: 24x, 2.1°

80mm f/6 Refractor

The 98 - 101 Aquarii Asterism / Aquarius / 98 Aqr / RA: 23h 22m 58s, Dec: -20° 06' 02" (2000)

Finder

Eyepiece: 18x, 2.7°

80 mm f/6 Refractor

The 88 - 86 Aquarii Asterism / Aquarius / 88 Aqr / RA: 23h 09m 27s, Dec: -21° 10' 21"

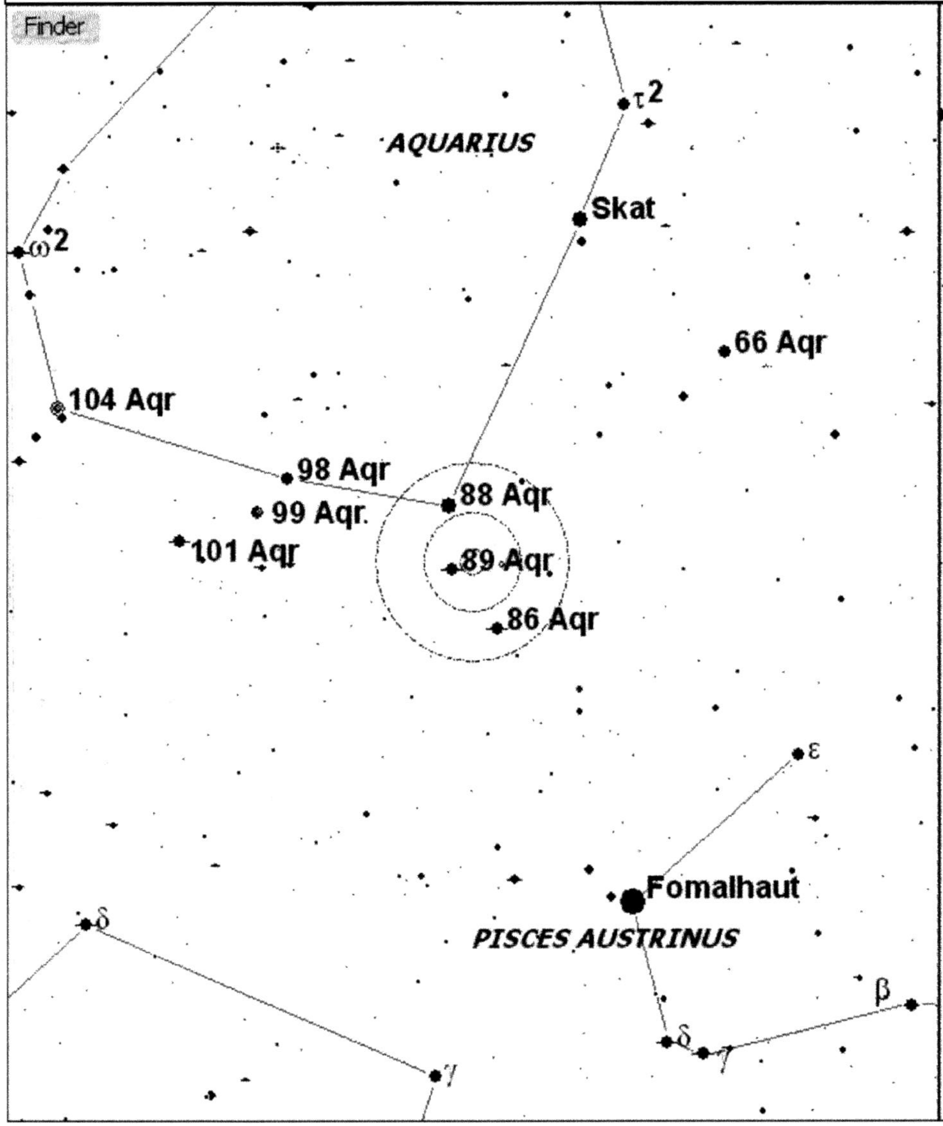

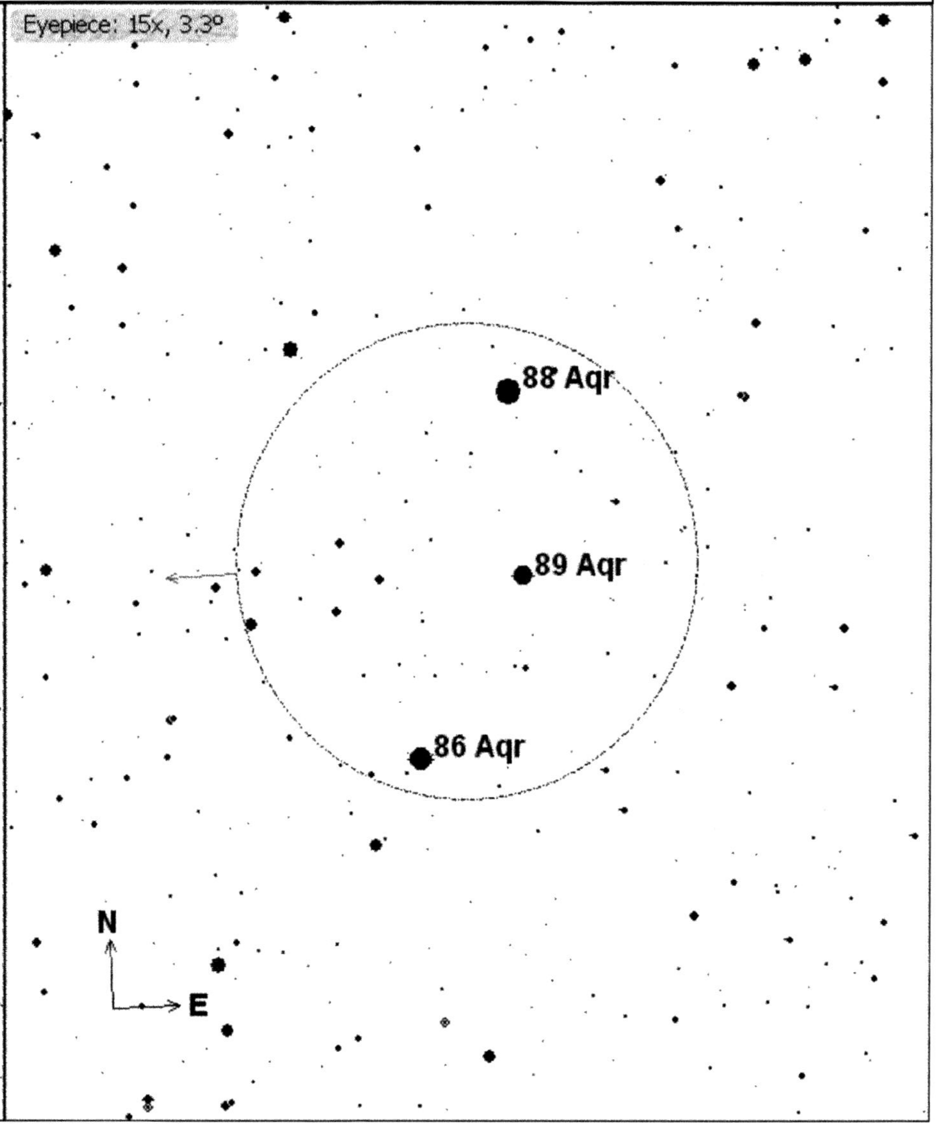

Finder

Eyepiece: 15x, 3.3°

80mm f/6 Refractor

The Omega (ω) Aquarii Asterism / Aquarius / Omega² Aqr / RA: 23h 42m 43s, Dec: -14° 32' 42"

Finder

Eyepiece: 24x, 2.1°

30 Psc
33 Psc
ψ2
ψ1
AQUARIUS
88 Aqr
98 Aqr
99 Aqr
89 Aqr
101 Aqr
104 Aqr
ω2
2 Cet
7 Cet

ω1
ω2
R Aqr

N E

80mm f/6 Refractor

The 103 - 108 Aquarii Asterism / Aquarius / 104 Aqr / RA: 23h 41m 46s, Dec: -17° 49' 00"

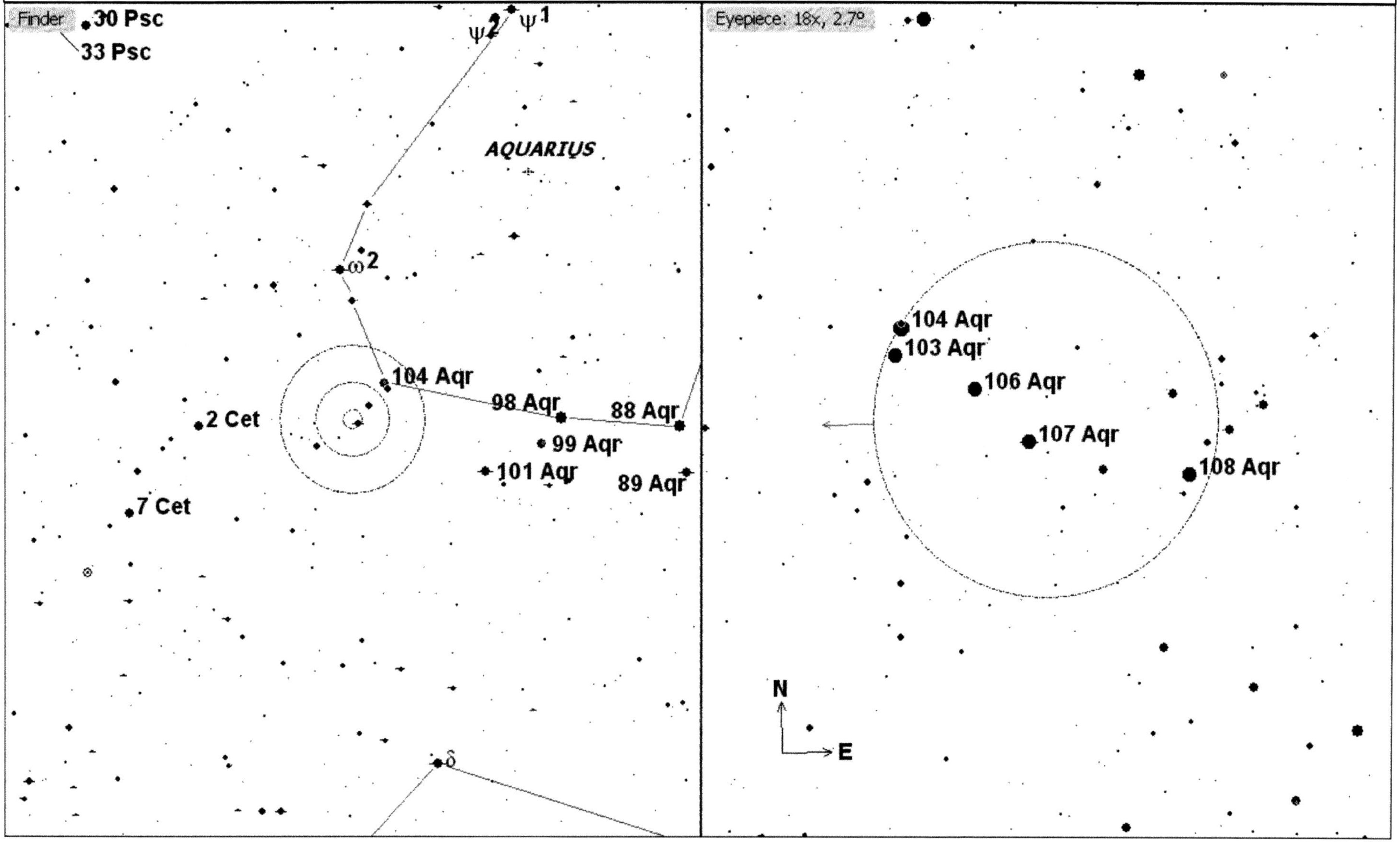

Eyepiece: 18x, 2.7°

80mm f/6 Refractor

Waters of Aquarius

Principal Stars Continued

100 Aqr   Single Star
aka HR 8932, HD 221357, SAO 191970
Magnitude: 6.24, F0
Distance: 260 +/- 18.0 ly

101 Aqr   Multiple Star System
aka HR 8939, HD 221565, SAO 191988
Magnitude: 4.72, A0
Distance: 320 +/- 24.0 ly
AB: 4.7+7.43 mag;
In 1991: PA 125° Sep 0.84"

103 Aqr   Single Star
aka HR 8980, HD 222547, SAO 165834
Magnitude: 5.33, K4III
Distance: 580 +/- 79.0 ly

104 Aqr   Multiple Star System
aka HR 8982, HD 222574, SAO 165836, HJ 5413
Magnitude: 4.83, G2I
Distance: 640 +/- 110.0 ly
AB: 4.82+7.9 mag, B=HD 222561;
In 1919: PA 6° Sep 120.10"
BC: 7.9+11.9 mag
In 1913: PA 76° Sep 90.20"

106 Aqr   Single Star
aka HR 8998, HD 222847, SAO 165854
Magnitude: 5.24, B9
Distance: 330 +/- 29.0 ly

107 Aqr   Multiple Star System
aka HR 9002, HD 223024, SAO 165867, H 24, ADS 16979
Magnitude: 5.28, F0
Distance: 210 +/- 16.0 ly
AB: 5.28+6.72 mag, B=HIP 117218
In 1991: PA 136° Sep 6.79"

Object data from SKYTOOLS™ 2 — used with permission

108 Aqr   Variable Star
aka ET Aqr, GCVS 783, HR 9031, HD 223640, SAO 165918
Type: ACV
Period: 3.730000 days
Magnitude: V 5.16 to V 5.21, A
Distance: 320 +/- 33.0 ly

Omega 2 Aqr   Multiple Star System
aka 105 Aqr, HR 8988, HD 222661, SAO 165842, BU 279, ADS 16944
Magnitude: 4.51, B9
Distance: 150 +/- 6.3 ly
AB: 4.49+10.6 mag
In 1875: PA 86° Sep 5.70"

Omega 1 Aqr   Single Star
aka 102 Aqr, HR 8968, HD 222345, SAO 165818
Magnitude: 4.97, A7IV
Distance: 130 +/- 4.7 ly

Psi 3 Aqr   Multiple Star System
aka 95 Aqr, HR 8865, HD 219832, SAO 146635, HO 199, ADS 16671
Magnitude: 4.99, A0
Distance: 250 +/- 19.0 ly
AB: 4.99+11.2 mag,
In 1959: PA 174° Sep 1.50"
AC: 4.99+13.3 mag,
In 1912: PA 230° Sep 130.10"

Psi 2 Aqr   Single Star
aka 93 Aqr, HR 8858, HD 219688, SAO 146620
Magnitude: 4.40, G8III
Distance: 320 +/- 33.0 ly

Psi 1 Aqr   Multiple Star System
aka 91 Aqr, HR 8841, HD 219449, SAO 146598, Struve 5012, ADS 16633, BU 1220
Magnitude: 4.25, K0III
Distance: 150 +/- 6.0 ly

Psi 1 Aqr

Component	Magnitudes	Date	Position Angle	Separation
AB	4.24+10.3	1836	312°	49.60"
BC	10.3+11.5	1958	105°	0.30"
AD	4.24+13.5	1924	274°	80.40"
BE	10.3+14.3	1924	341°	19.70"

The Psi[1,2,3] (ψ[1,2,3]) Aquarii Asterism / Aquarius / Psi[1] Aqr / RA: 23h 15m 54s, Dec: -09° 05' 16"

Eyepiece: 24x, 2.1°

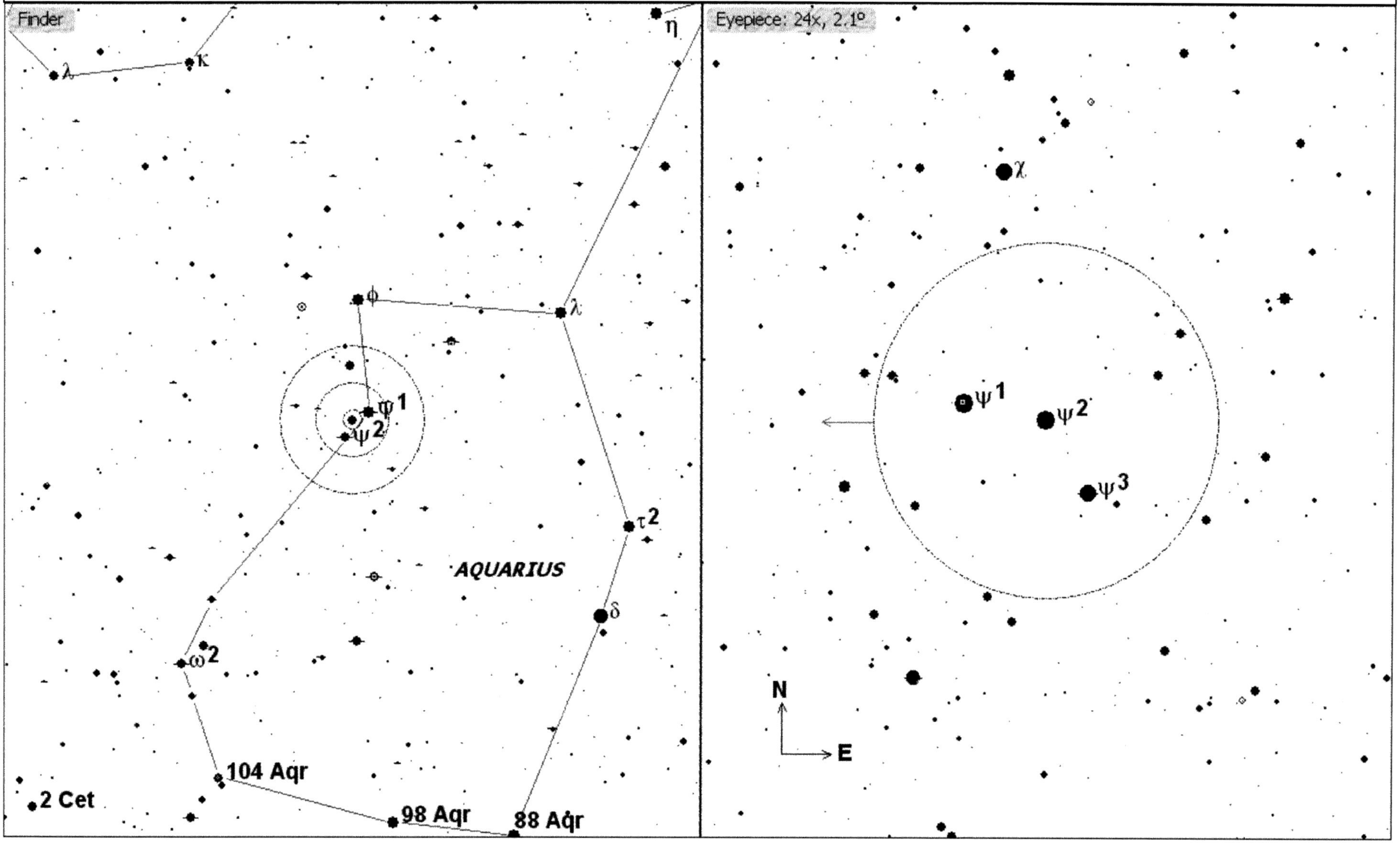

80mm f/6 Refractor

Name	**M 73**
Location	RA: 20h 58m 57s, Dec: -12° 38' 30" (2000) in Aquarius
Compare with	15 Cephei, NGC 2017
Found With	Telescope
Description	M 73 is the classic asterism. The stars are described as "not related" or "just an asterism" in astronomical literature. However belittled by serious writers, M73 is guaranteed a place in observer's logbooks by its name. The asterism is tiny: is has a diameter of just over one arc-minute. Compare with multiple stars 15 Cephei and NGC 2017. A detailed analysis of NGC 2017 can be found in the *Night Sky Observer's Guide Vol. 1*; see bibliography on page 4.

```
        •TYC 05778-0509 1 (113)
                              •TYC 05778-0594 1 (117)
                 •TYC 05778-0492 1 (119)


        •TYC 05778-0802 1 (104)
```

Principal Stars	
TYC 05778-0594 1 Single Star aka SI 1955780 Magnitude: 11.72	***TYC 05778-0802 1** Single Star aka PPM 722368 , SI 598121 Magnitude: 10.38
TYC 05778-0492 1 Single Star aka SI 1955775 Magnitude: 11.87	**TYC 05778-0509 1** Single Star aka SI 598102 Magnitude: 11.27

Object data from SKYTOOLS™ 2 — used with permission.

M73 / Aquarius / TYC 05778-0802 1 / RA: 20h 58m 57s, Dec: -12° 38' 30"

Naked-Eye

Finder

Eyepiece: 114x, 26.3'

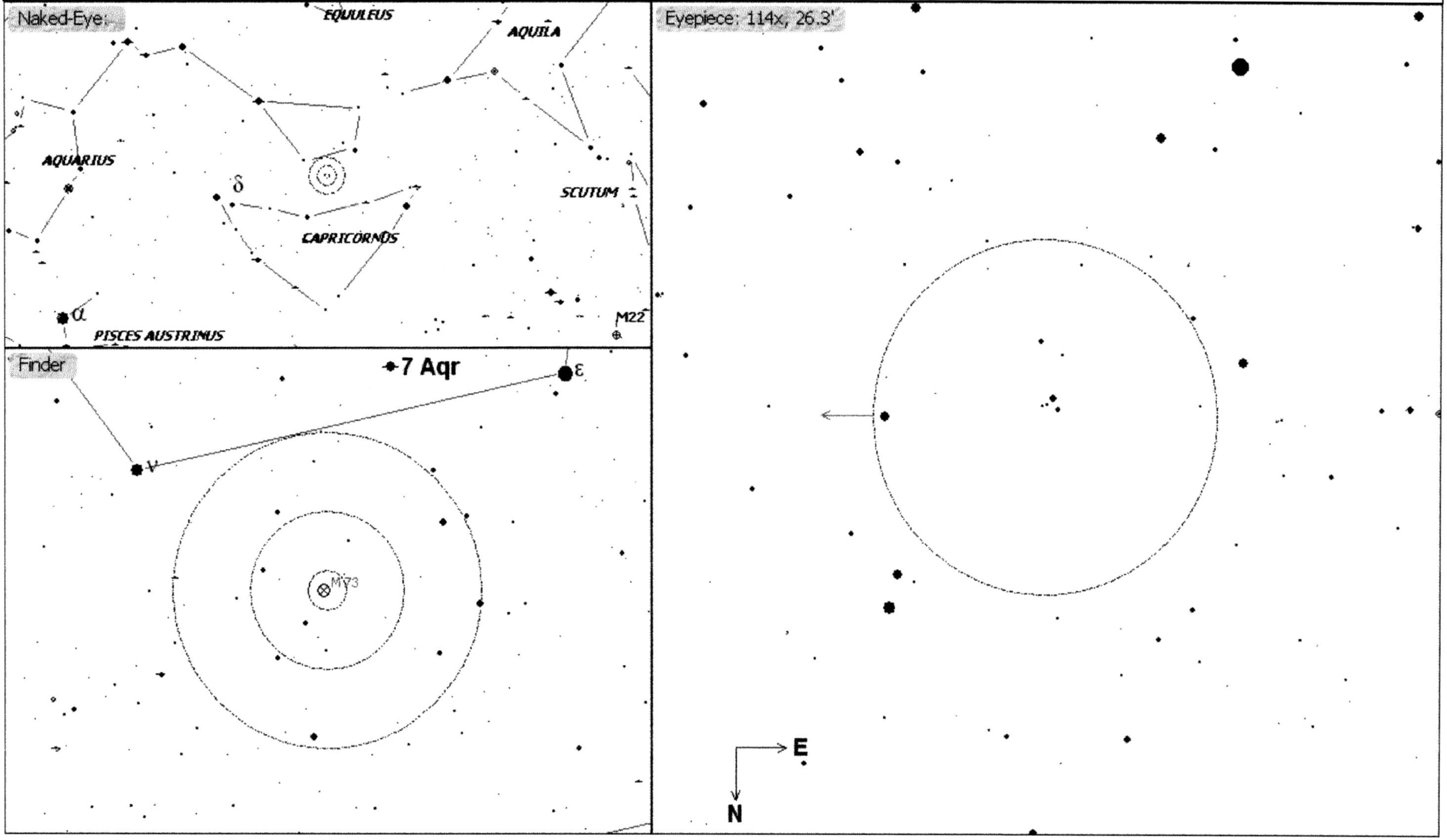

6" f8 Reflector

Name	**The Water Jar Asterism**
Location	RA: 22h 28m 50s, Dec. -00° 01' 12" in Aquarius
Found With	Naked Eye
Description	This is the famous asterism of Aquarius. Zeta is a favorite of double star observers. The Water Jar is the source of the waters of eastern Aquarius.

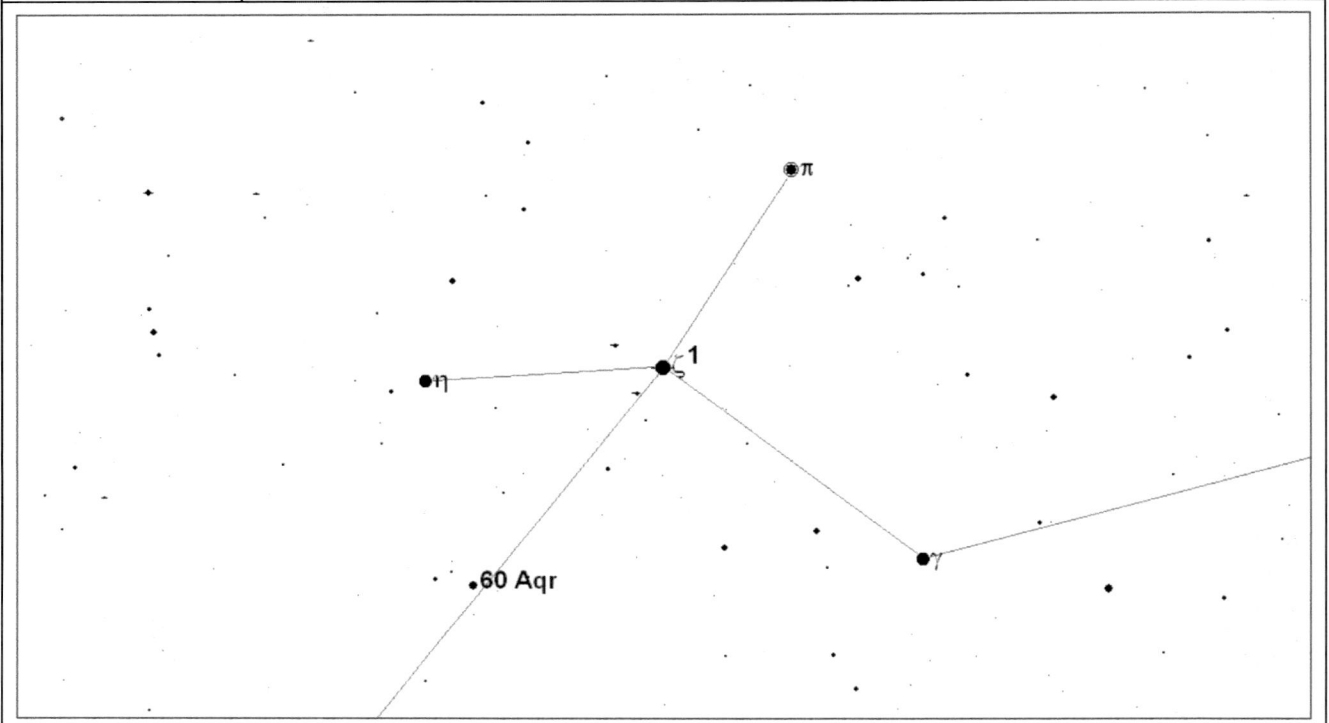

Principal Stars

Gamma γ Aqr Multiple Star System
aka Sadalachbia , 48 Aqr, HR 8518,
HD 212061, ADS 15864, SAO 146044
Magnitude: 3.84, A0
Distance: 160 +/- 13.0 ly
AB: 3.86+12.2 mag,
In 1959: PA 140° Sep 37.40"

Pi π Aqr Variable Star
aka 52 Aqr, Pi Aqr, GCVS 799, HR 8539,
HD 212571, SAO 127520
Mag: V 4.42 to V 4.70, B1 Shell Star
Distance: 1100 +/- 260.0 ly

*Zeta ζ Aqr Multiple Star System
aka 55 Aqr, HR 8558, HD 213051,
SAO 146107, Struve 2909, ADS 15971
Magnitude: 3.65, F2
Distance: 100 +/- 4.9 ly
AB: 3.65+4.6 mag, B=Zeta 2 Aqr;
PA 178° Sep 2.06" (2004.7)

Eta η Aqr Single Star
aka 62 Aqr, HR 8597, HD 213998,
SAO 146181
Magnitude: 4.00, B9IV
Distance: 180 +/- 8.5 ly

Object data from SKYTOOLS™ 2 — used with permission.

The Water Jar Asterism / Aquarius / Zeta (ζ) Aqr / RA: 22h 28m 50s, Dec: -00° 01' 12"

Eyepiece: 14x, 5.0°

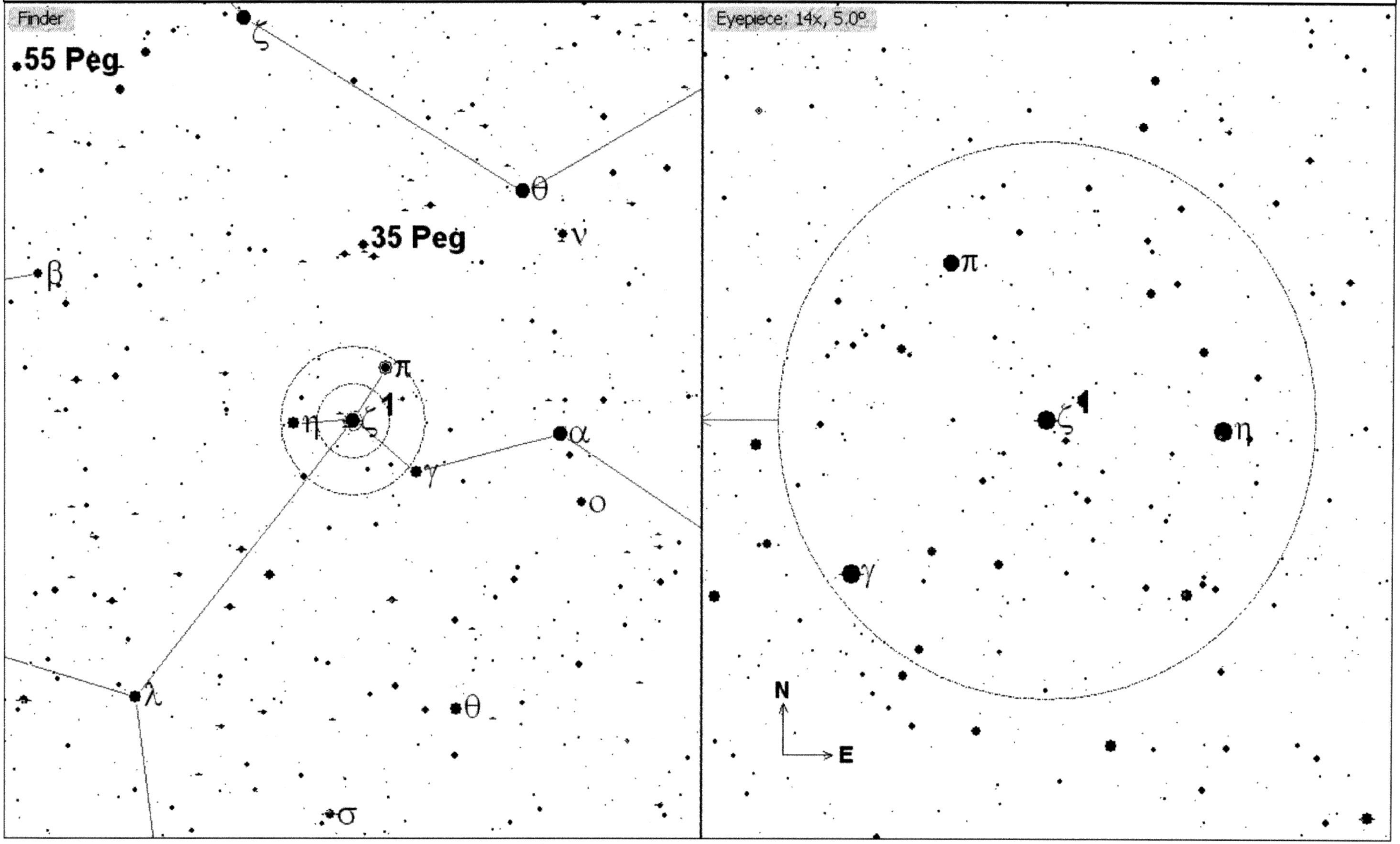

80mm f/6 Refractor

Name	**The Xi (ξ) Arietis Asterism**
Location	RA: 02h 24m 49s, Dec: +10° 36' 38" (2000) in Aries
Found With	16 x 70 Binoculars
Description	This is a conspicuous binocular group just above the head of Cetus.

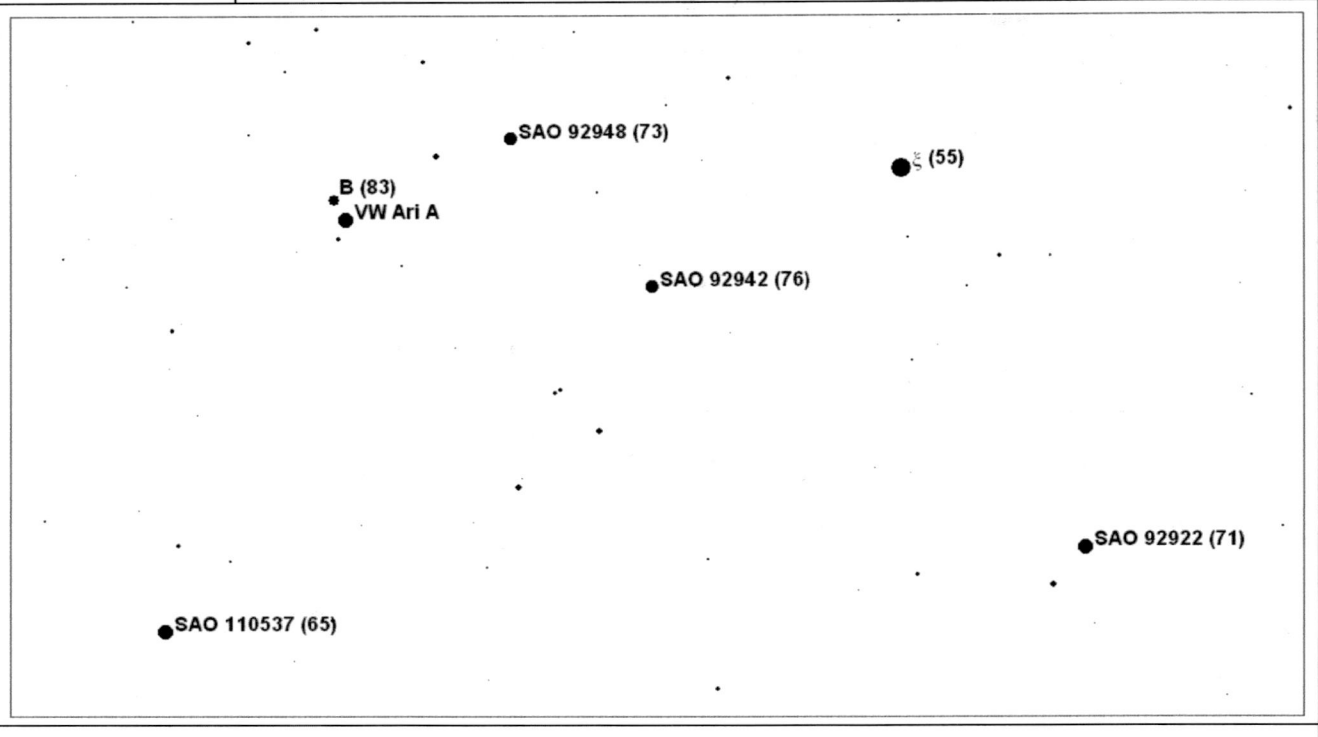

Principal Stars

SAO 92922 Single Star
aka HD 14866, PPM 118189, HIP 11194
Magnitude: 7.09, K0
Distance: 420 +/- 52.0 ly

***Xi (ξ) Ari   Single Star**
aka 24 Ari, HR 702, HD 14951, SAO 92932
Magnitude: 5.48, B7IV
Distance: 600 +/- 120.0 ly

SAO 92942   Single Star
aka HD 15042, PPM 118219, HIP 11319
Magnitude: 7.61
Magnitude: 7.61, B9
Distance: 990 +/- 290.0 ly

VW Ari   Multiple Star System
aka GCVS 3024, HD 15165, SAO 92952,
Otto Struve 27
Variability: Type: SXPHE
Mag: V 6.64 to V 6.76, A5
Period: 0.149000 days
Distance: 380 +/- 40.0 ly
AB: 6.7+8.3 mag, B=HD 15164
In 1875: PA 31° Sep 73.80"
AC: 6.7+11.3 mag,
 In 1898: PA 155° Sep 62.30"

SAO 110537   Single Star
aka HD 15228, PPM 118239, HIP 11427
Magnitude: 6.45, F5
Distance: 120 +/- 3.8 ly

Object data from SKYTOOLS™ 2 — used with permission.

The Xi (ξ) Arietis Asterism / Aries / RA: 02h 24m 49s, Dec: +10° 36' 38"

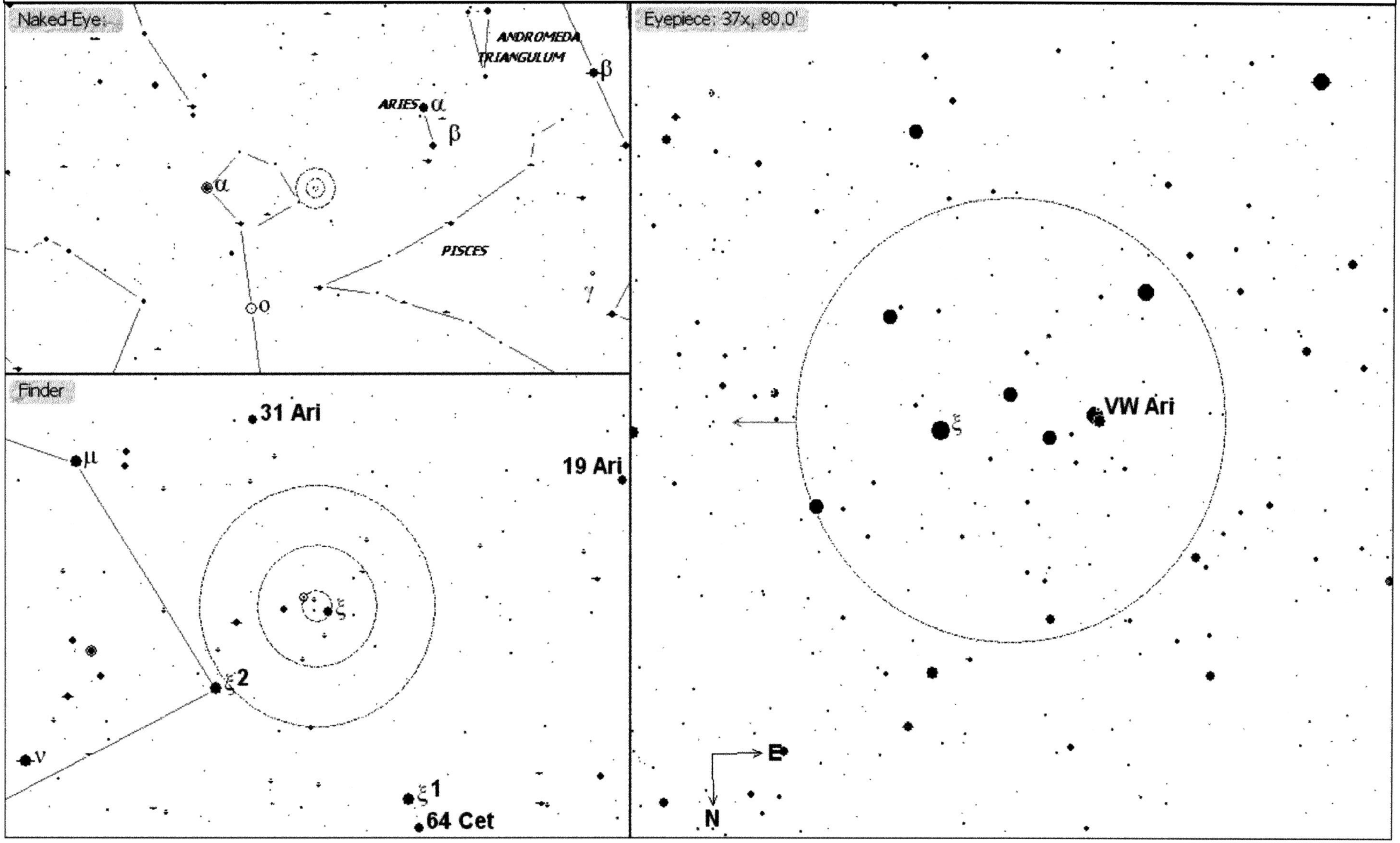

Naked-Eye

Finder

- 19 Ari
- 31 Ari
- 64 Cet
- ξ1
- ξ2
- μ
- ν

Eyepiece: 37x, 80.0'

VW Ari
ξ

N / E

6" f/8 Reflector

25

Name	**The "Island of Auriga" Asterism**
Location	RA: 05h 18m 11s, Dec: +33° 22' 18" (2000)
Colorful Stars	16 Aur
Found With	Direct Vision
Description	This asterism appears as a conspicuous "island" in the river of the Milky Way. The distance between 14 and 19 Aur is about 1.5 degrees.

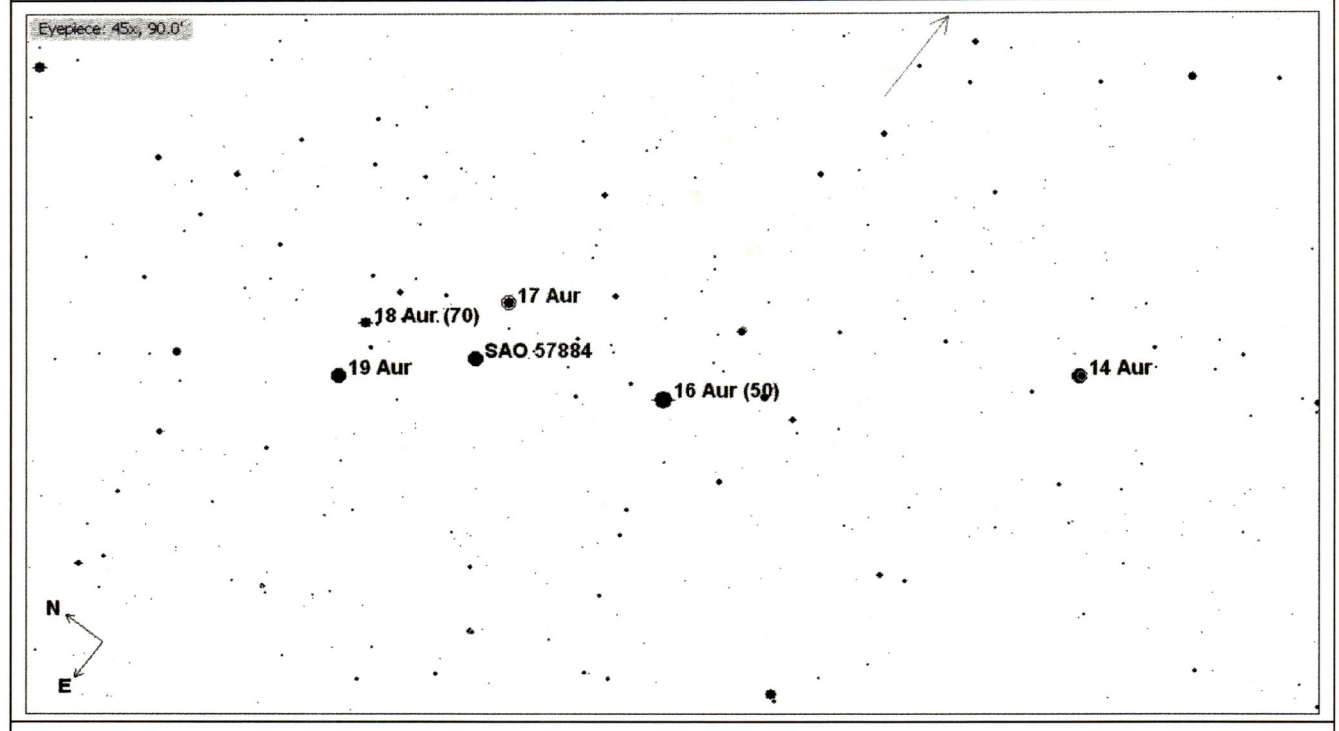

Principal Stars

14 Aur Multiple Star System
aka KW Aur, GCVS 3283, HR 1706, HD 33959, SAO 57799, Struve 653, ADS 3824
Mag: V 4.95 to V 5.08, A2
Distance: 270 +/- 26.0 ly
AB: 5.01+11.1 mag;
In 1909: PA 352° Sep 11.1"
AC: 5.01+7.86 mag, C=SAO 57798
In 1991: PA 225° Sep 14.29"
AD: 5.01+10.4 mag; In 1908: PA 321° Sep 184"

*16 Aur Multiple Star System
aka HR 1726, HD 34334,
SAO 57853, ADS 3872
Magnitude: 4.54, K0
Distance: 230 +/- 23.0 ly
AB: 4.54+10.6 mag;
In 1848: PA 56° Sep 4.20"

SAO 57884 Variable Star
aka IQ Aur, GCVS 3261, HR 1732, HD 34452
Mag: V 5.35 to V 5.43, A0
Distance: 450 +/- 43.0 ly

17 Aur Variable Star
aka AR Aur, GCVS 3098, HD 34364,
SAO 57858
Mag: V 6.15 to V 6.82, B9.5
Period: 4.134695 days
Distance: 400 +/- 38.0 ly

18 Aur Multiple Star System
aka HR 1734, HD 34499,
SAO 57893, ADS 3893
Magnitude: 6.49, A7
Distance: 240 +/- 14.0 ly
AB: 6.49+11.8 mag;
In 1883: PA 165° Sep 4.10"

19 Aur Variable Star
aka HR 1740, HD 34578, SAO 57906
Mag: V 5.03 to V 5.09, A5II
Distance: 3100 +/- 2200.0

Object data from SKYTOOLS™ 2 — used with permission

The Island of Auriga Asterism / Auriga / 16 Aur / RA: 05h 18m 11s, Dec: +33° 22' 18"

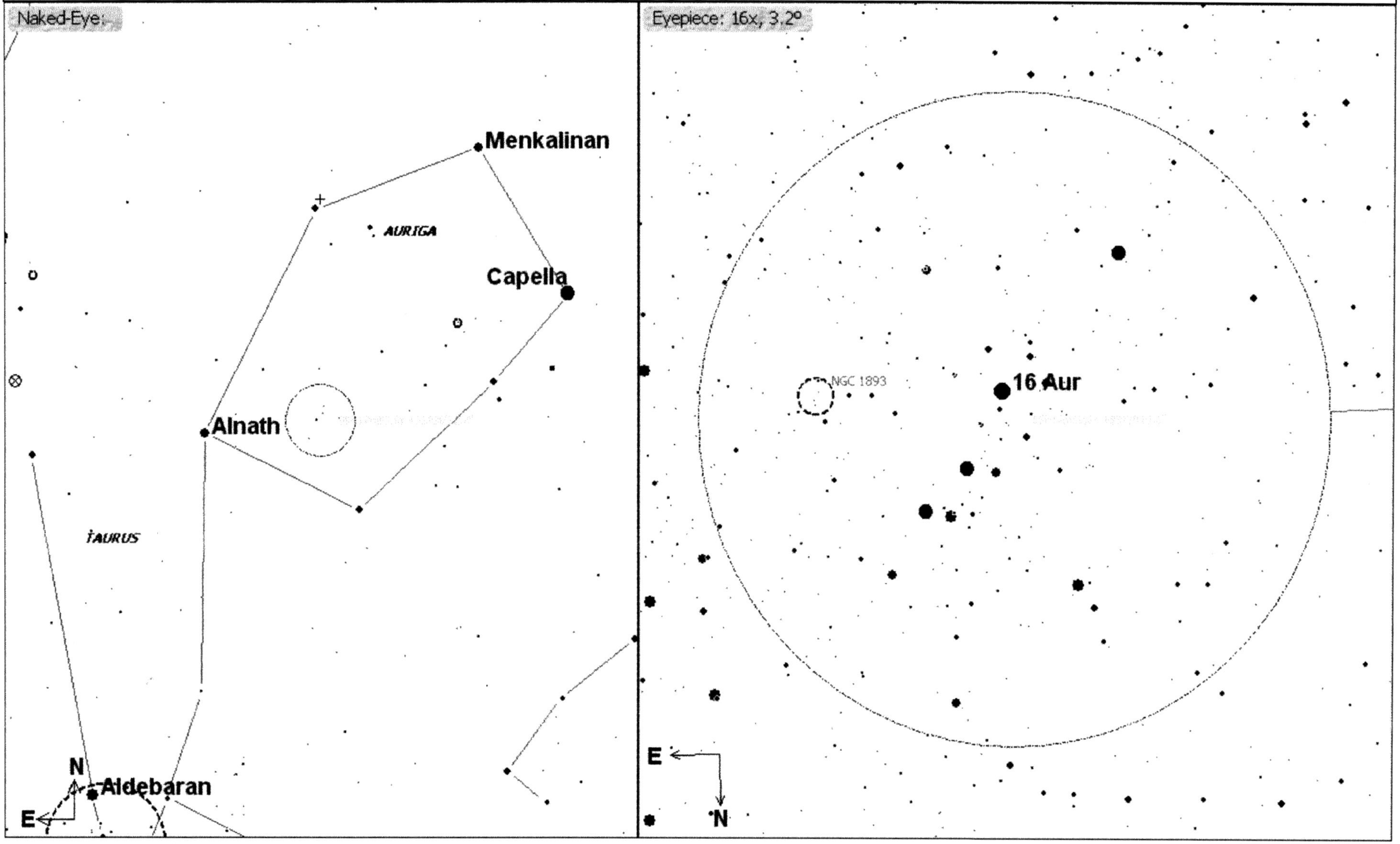

Naked-Eye

Eyepiece: 16x, 3.2°

16 Aur

NGC 1893

Menkalinan
Capella
AURIGA
Alnath
TAURUS
Aldebaran

N
E

16 x 70 Binoculars

Name	**The "Kids" Asterism**
Location	RA: 05h 06m 31s, Dec: +41° 14' 04" (2000) in Auriga
Colorful Stars	ζ Zeta, η Eta Aur
Found With	Direct Vision
Description	This 3 star asterism is known to astronomers as the "Kids", the young goats that the charioteer Auriga carries under his arm. It is a bright, distinctive triangle southwest of Capella. The pair of Zeta and Eta contrast sharply in color when seen in binoculars or low-power telescope: Zeta is a warm orange, Eta an electric blue. Epsilon is a famous eclipsing variable star.

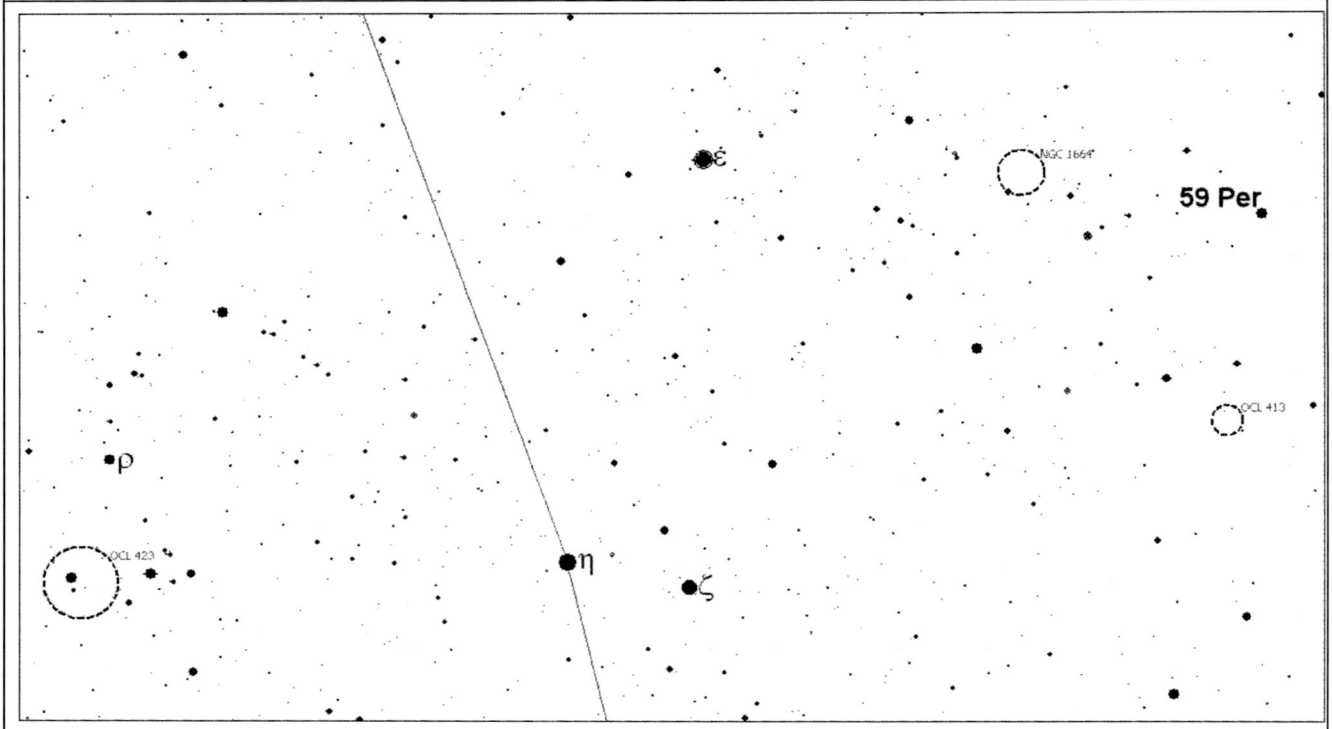

Principal Stars

Zeta Aur Variable Star
aka 8 Aur, GCVS 3390, HD32068, SAO39966
Mag: V 3.70 to V 3.97, K5
Distance: 790 +/- 150.0 ly

*Eta Aur Variable Star
aka 10 Aur, HR 1641, HD 32630, SAO 40026
Mag: V 3.16 to V 3.19, B3
Distance: 220 +/- 11.0 ly

Epsilon Aur Multiple Star System
aka, 7 Aur, GCVS 3389, HR 1605, HD 31964, SAO 39955, BU 554, ADS 3605
Mag: V 2.92 to V 3.83, F2I, Distance: 2000 +/- 1500.0 ly

Component	Magnitudes	Date	Position Angle	Separation	Other Name
AB	3.03+14	1878	PA 224°	Sep 21.20"	
AC	3.03+11.7	1878	PA 275°	Sep 43.00"	
AD	3.03+12	1879	PA 317°	Sep 46.20"	
AE	3.03+9.2	1913	PA 48°	Sep 207.60"	E=SAO 39960

Object data from SKYTOOLS™ 2 — used with permission.

The "Kids" Asterism / Auriga / Eta η Aur / RA: 05h 06 m31s, Dec: +41° 14' 04" (2000)

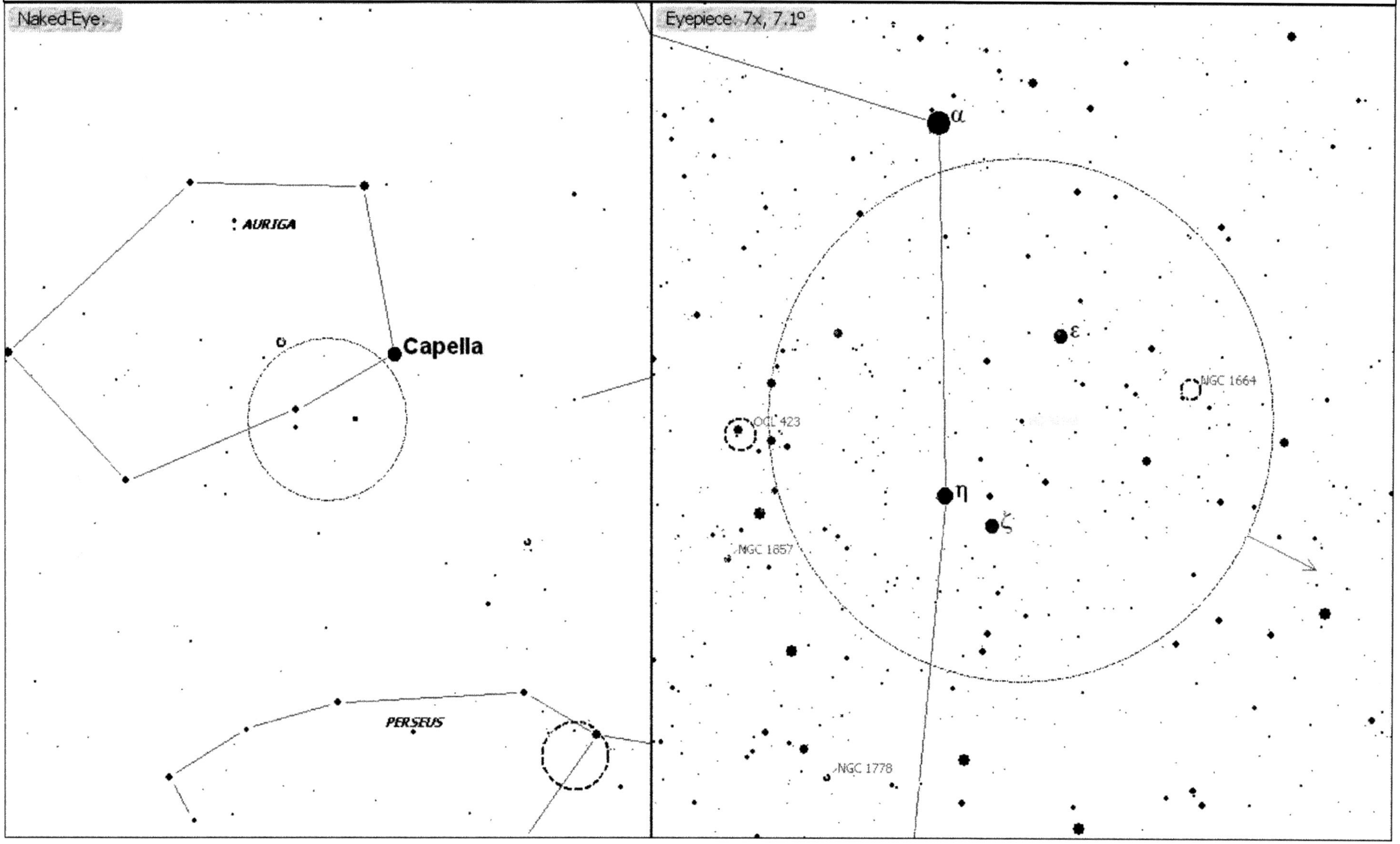

Naked-Eye:

Eyepiece: 7x, 7.1°

7x50 Binoculars

Name	**The "Reverse Kids" Asterism**
Location	RA: 05h 51m 29s, Dec: +39° 08' 55" (2000) in Auriga
Colorful Stars	Tau τ, Nu ν, Upsilon υ Aurigae
Found With	Direct Vision
Description	This 3 star asterism forms a narrow triangle that points south. Compare this asterism to the well-known "Kids" asterism in western Auriga, a narrow triangle that points north. The separation between Tau and Nu is about half of a degree and from Tau to Upsilon is 2 degrees.

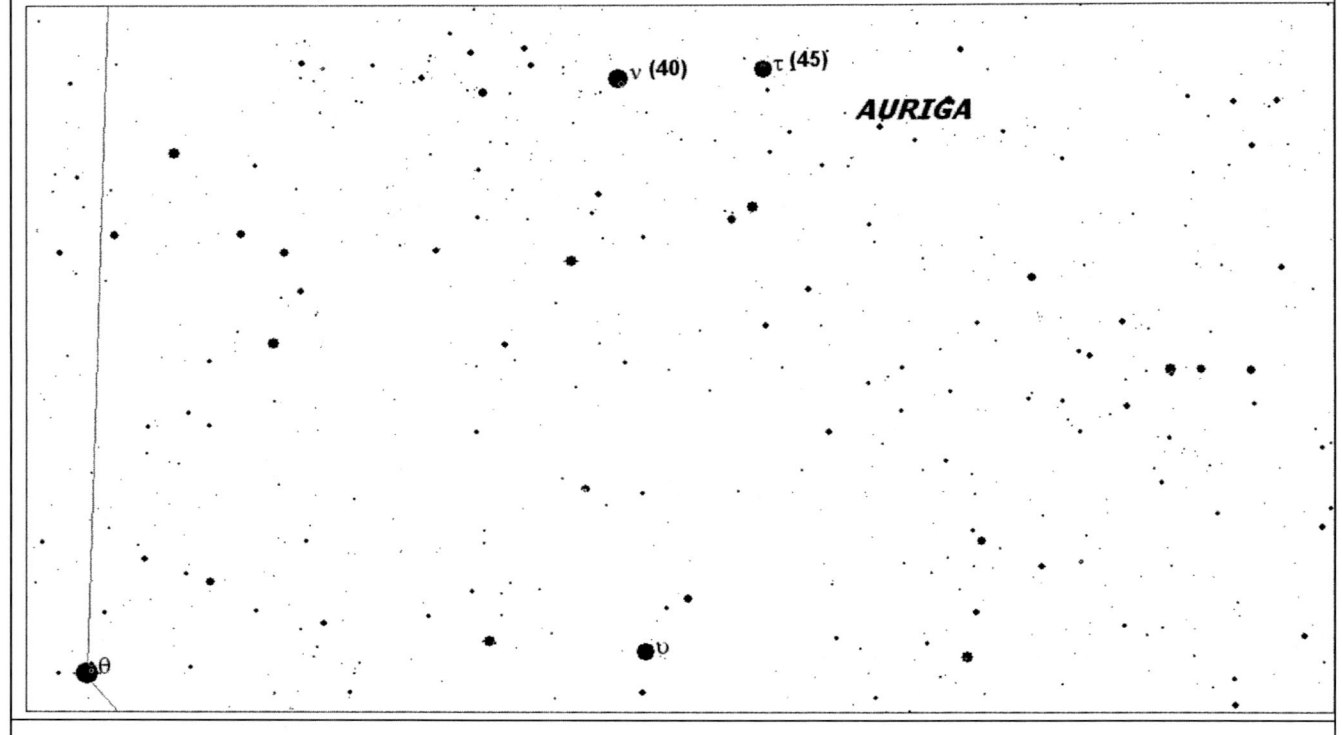

Principal Stars

Tau (τ) Aur Multiple Star System
aka 29 Aur, HR 1995, HD 38656,
SAO 58465, BU 192, ADS 4398
Magnitude: 4.53, G8III
Distance: 210 +/- 11.0 ly
AB: 4.51+11.6 mag
In 1877: PA 352° Sep 39.40"
AC: 4.51+11.6 mag
In 1926: PA 35° Sep 49.60"

*Nu (ν) Aur Multiple Star System
aka 32 Aur, HD 39003,
SAO 58502, ADS 4440
Magnitude: 3.97, G9.5 III
Distance: 210 +/- 12.0 ly
AB: 3.97+9.5 mag, H 90
In 1878: PA 206° Sep 54.60"

Upsilon (υ) Aur Variable Star
aka 31 Aur, HR 2011, HD 38944,
SAO 58496
Mag: V 4.73 to V 4.83, M0
Distance: 480 +/- 53.0 ly

Object data from SKYTOOLS™ 2 — used with permission.

The "Reverse Kids" Asterism / Auriga / Nu (ν) Aur / RA: 05h 51m 29s, Dec: +39° 08' 55"

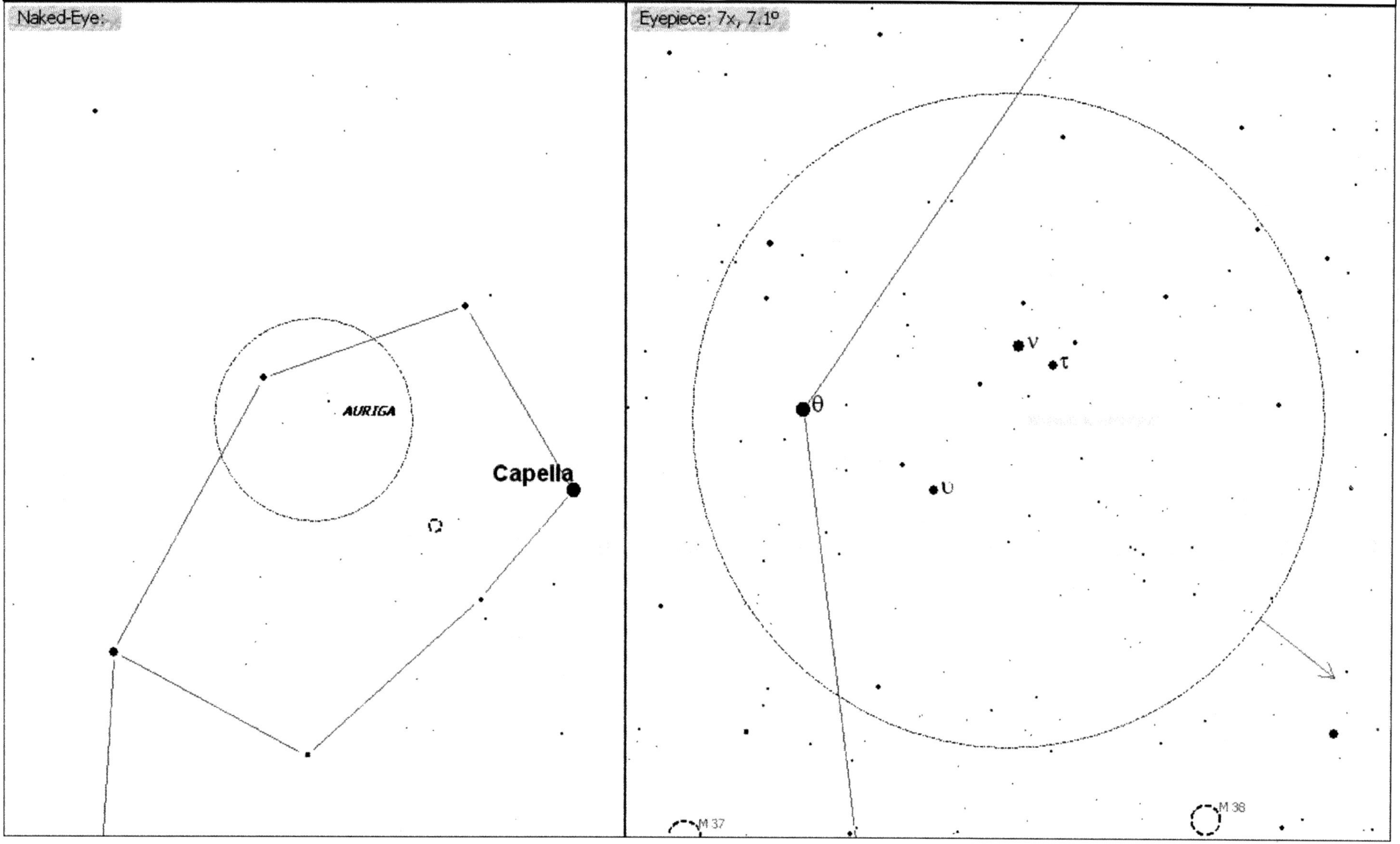

Naked-Eye:

Eyepiece: 7×, 7.1°

7×50 Binoculars

Name	**The Kappa - Iota Boötes Asterism**
Location	RA: 14h 16m 10.0s, Dec: +51° 22' 02" (2000)
Colorful Stars	SAO 29058, SAO 29062, SAO 26066, SAO 29089
Found With	Direct Vision
Description	The Kappa - Iota Boötis region appears as a fuzzy spot to the naked eye. Binoculars or a low-power telescope reveal a lovely gathering of stars. Field shown is 2° wide, 1°15' tall.

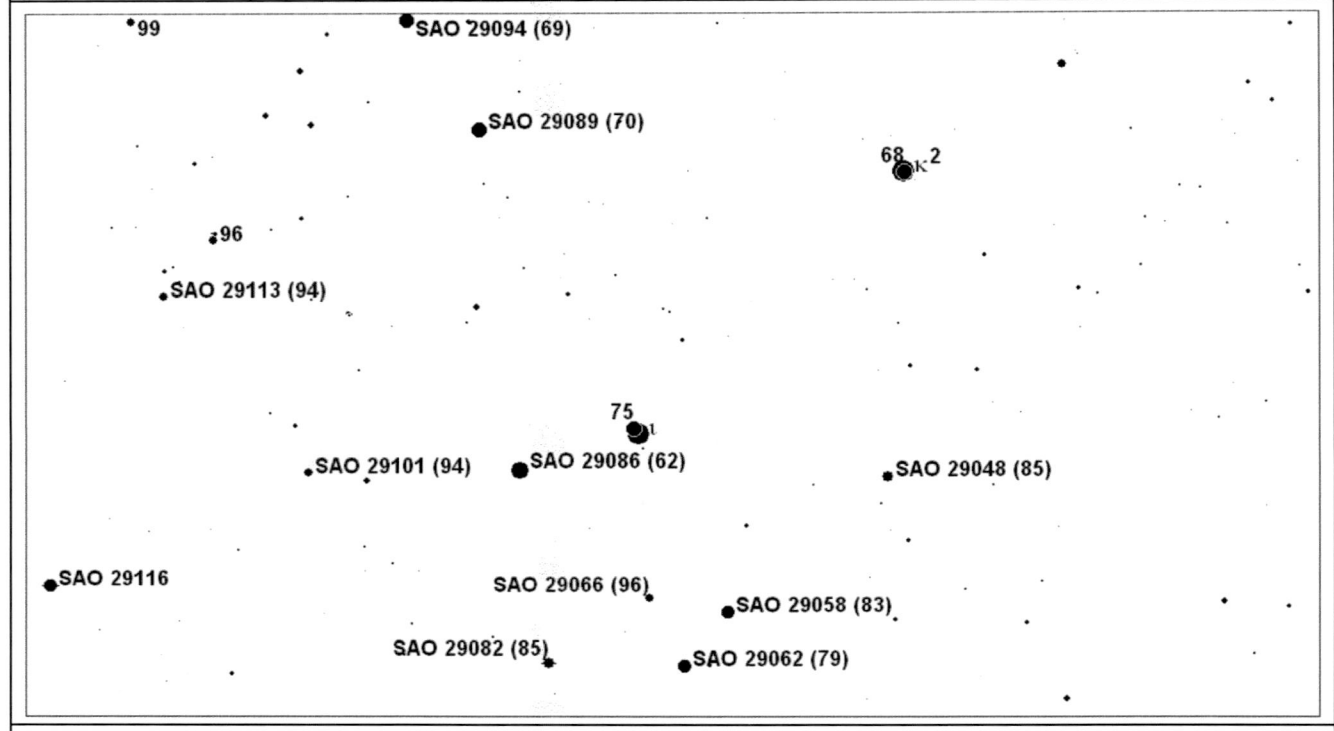

Principal Stars

Kappa-2 Boö Multiple Star System
aka, 17 Boo, Struve 1821, ADS 9173, HR 5329, HD 124675, SAO 29046
Magnitude: 4.41, A5, Distance: 160 +/- 6.1 ly
AB: 4.53+6.81 mag, B=Kappa-1 Boo
In 1991: PA 236° Sep 13.50"

SAO 29048 Single Star
aka HD 124693
Magnitude: 8.47, F8
Distance: 1000 +/- 260.0 ly

SAO 29058 Single Star
aka HD 124984, PPM 34418
Magnitude: 8.33, K2
Distance: 1300 +/- 420.0 ly

SAO 29062 Single Star
aka HD 125075, PPM 34422
Magnitude: 7.93, K0
Distance: 1200 +/- 360.0 ly

SAO 29066 Single Star
aka HD 234118, PPM 34428
Magnitude: 9.60, K2

Object data from SKYTOOLS™ 2 — used with permission.

*Iota Boö Multiple Star System
aka 21 Boo, HR 5350, HD 125161, ADS 9198, SAO 29071, Struve 4026
Magnitude: 4.75, A9; Distance: 97 +/- 1.6 ly
AB: 4.75+7.5 mag, B=HD 234121
In 1836: PA 33° Sep 38.50"
AC: 4.75+12.6 mag
In 1911: PA 197° Sep 85.90"

SAO 29082 Multiple Star System
aka HD 125307, PPM 34442, ADS 9206
Magnitude: 8.48, G5
Distance: 270 +/- 36.0 ly
AB: 8.48+9.98 mag, In 1991: PA 103° Sep 0.64"

SAO 29086 Single Star
aka HR 5360 HD 125349,, PPM 34445
Magnitude: 6.19, A0.5 IV
Distance: 330 +/- 20.0 ly

SAO 29089 Single Star
aka HD 125435, PPM 34450
Magnitude: 7.00, K2

SAO 29094 Single Star
aka HD 125557, PPM 34454
Magnitude: 6.91, A2
Distance: 330 +/- 21.0 ly

The Kappa - Iota (κ - ι) Boötis Asterism / Boötes / Iota Boö / RA: 14h 16m 10.0s, Dec: +51° 22' 02"

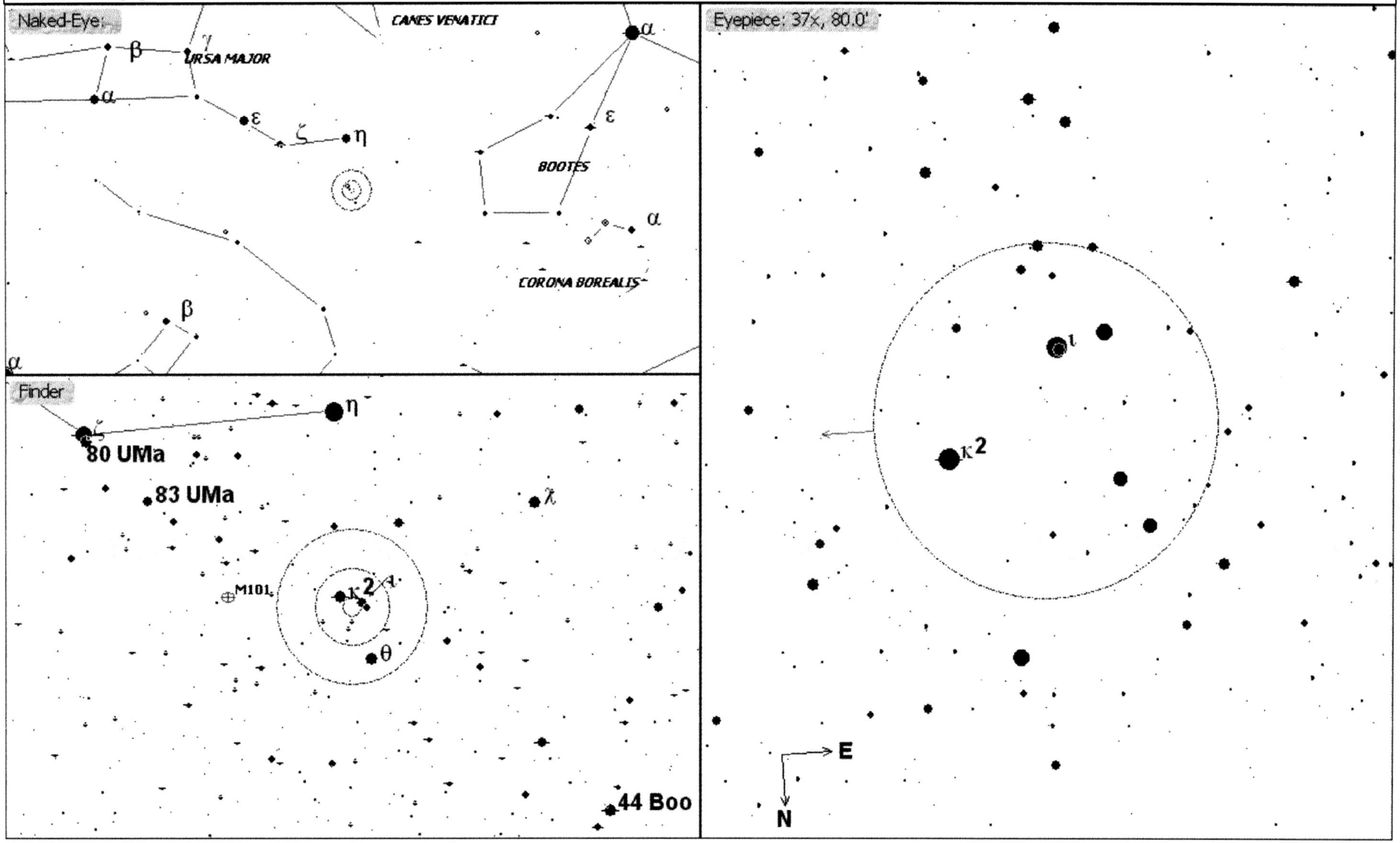

6" f/8 Reflector

Name	**The Struve 1835 Asterism**
Location	RA: 14h 23m 23s, Dec: +08° 26' 48" (2000) in Boötes
Colorful Stars	SAO 120436
Found With	Direct Vision
Description	This trio appears as a fuzzy spot to the naked eye. Σ1835 has no Flamsteed number, but is brighter than 14, 15, and 18 Boö. Field below is 1° by 37'.

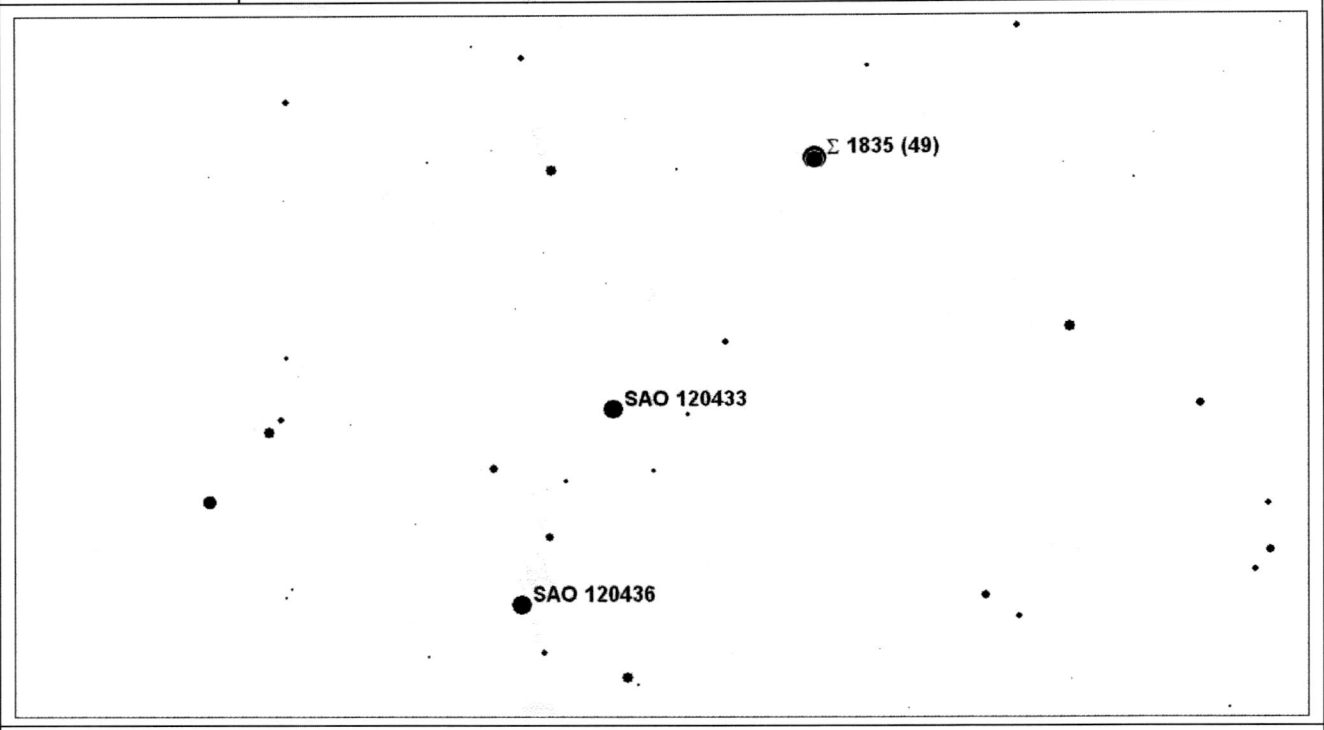

Principal Stars

*Struve 1835 Multiple Star System
aka HR 5386, HD 126129,
HIP 7032, ADS 9247
Magnitude: 4.86
Distance: 210 +/- 13.0 ly
AB: 4.86+7.03 mag
B=HR 5385
In 1991: PA 194° Sep 6.25"

SAO 120433 Variable Star
aka HR 5388, HD 126200, PPM 160594
Magnitude: 5.94, A3
Distance: 380 +/- 37.0 ly

SAO 120436 Variable Star
aka HR 5394, HD 126271, PPM 160598
Magnitude: 6.19, K2
Distance: 360 +/- 37.0 ly
Variability: Mag: V 6.16 to V 6.20

Object data from SKYTOOLS™ 2 — used with permission.

The Struve 1835 Asterism / Boötes / Σ1835 / RA: 14h 23m 23s, Dec: +08° 26' 48"

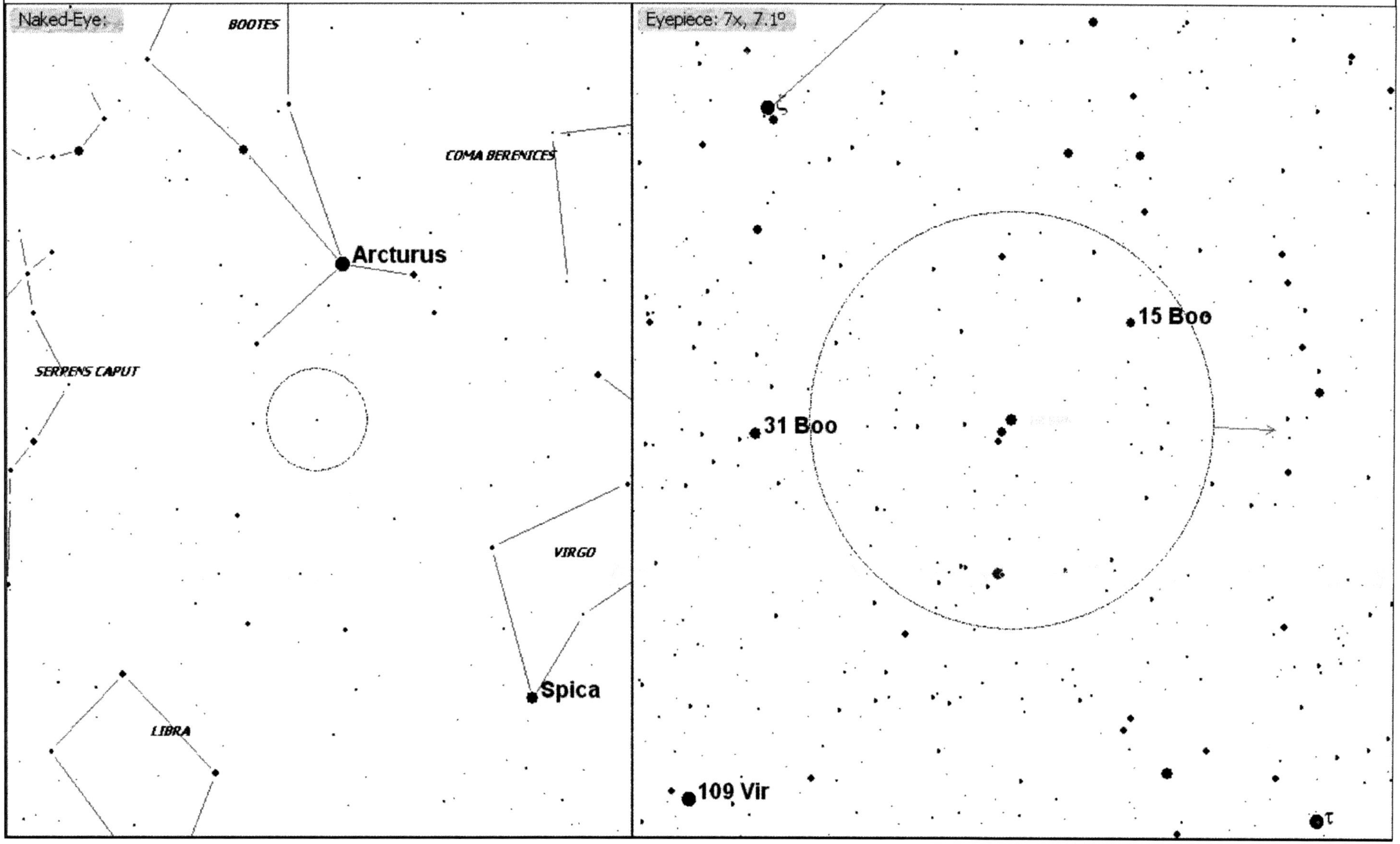

Naked-Eye:

Eyepiece: 7x, 7.1°

7x50 Binoculars

Name	**The Camelopardalis Asterism**
Location	RA: 12h 12m 12s, Dec: +77° 36' 58" (2000) in Camelopardalis
Colorful Stars	SAO 7500, SAO 7513
Found With	6 inch Dobsonian
Description	I first saw this asterism while looking for double star 32 Cam (Struve 1694). I checked my star atlases, but no open cluster is listed in this area. This asterism is visible to the naked eye as a fuzzy spot near Polaris. The two outliers, SAO 7500 and 7497, I consider to be part of this asterism. This is a true asterism - stars of varying distances appearing very close together. This field is 2° wide by 1°15' tall.

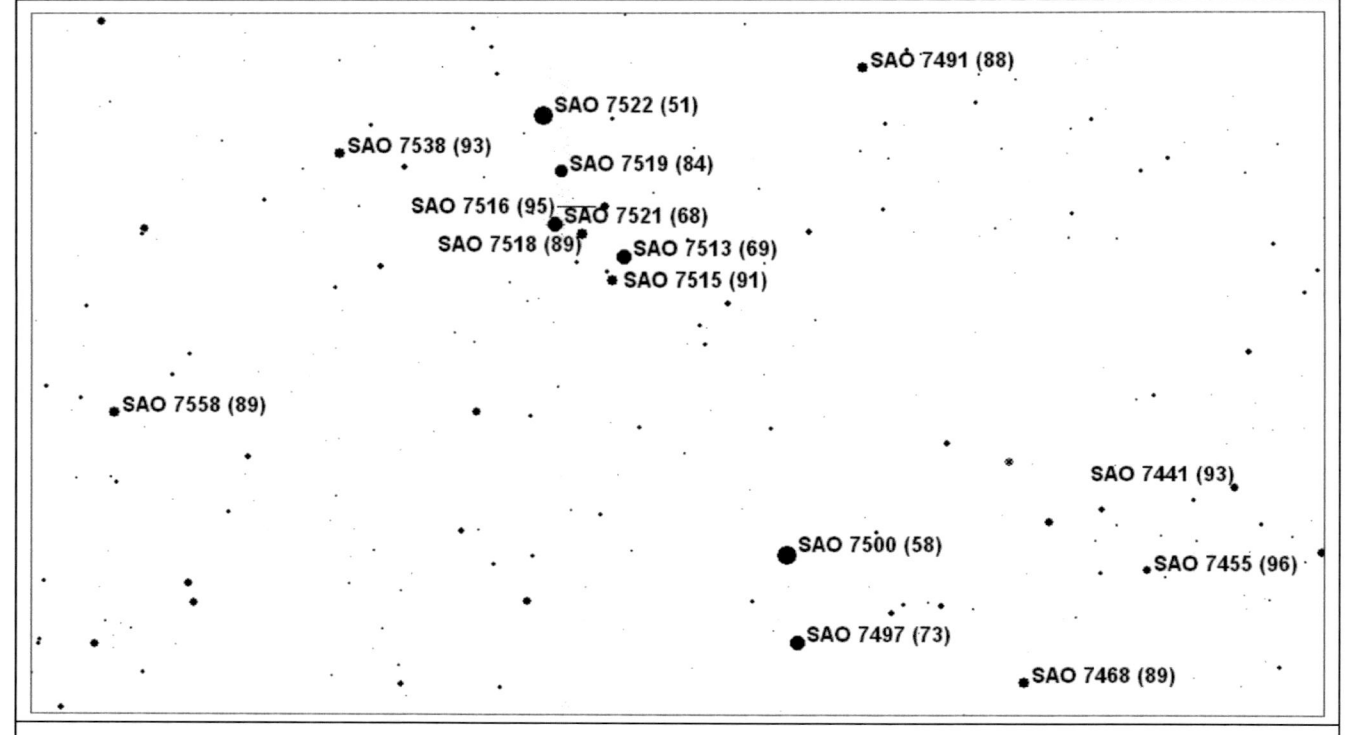

Principal Stars

SAO 7497 Single Star
aka HD 104954, PPM 8079, HIP 58924
Magnitude: 7.31, F5
Distance: 190 +/- 7.3 ly

SAO 7500 Single Star
aka HD 104985, HR 4609, PPM 8082
Magnitude: 5.78, K0
Distance: 330 +/- 18.0 ly

SAO 7513 Single Star
aka HD 105717, PPM 8096, HIP 59294
Magnitude: 6.87, G5
Distance: 270 +/- 13.0 ly

SAO 7515 Single Star
aka PPM 8098, BD +78 00407
Magnitude: 9.09

SAO 7516 Single Star
aka PPM 8099, BD +78 00408
Magnitude: 9.54

SAO 7518 Single Star
aka HD 105895, PPM 8103, BD +78
Magnitude: 8.86, A5

SAO 7521 Multiple Star System
aka HD 106053, PPM 8105, HIP 59463
Magnitude: 6.76, A0
Distance: 710 +/- 96.0 ly
AB: 6.76+10.89 mag
In 1991: PA 277° Sep 0.64"

SAO 7519 Single Star
aka PPM 8104, BD +78 00410, SI 493917
Magnitude: 8.43

*SAO 7522 Single Star
aka HR 4646, HD 106112, PPM 8107
Magnitude: 5.14, A5
Distance: 110 +/- 1.9 ly

Object data from SKYTOOLS™ 2 — used with permission

The Camelopardalis Asterism / Camelopardalis / SAO 7522 / RA: 12h 12m 12s, Dec: +77° 36' 58" (2000)

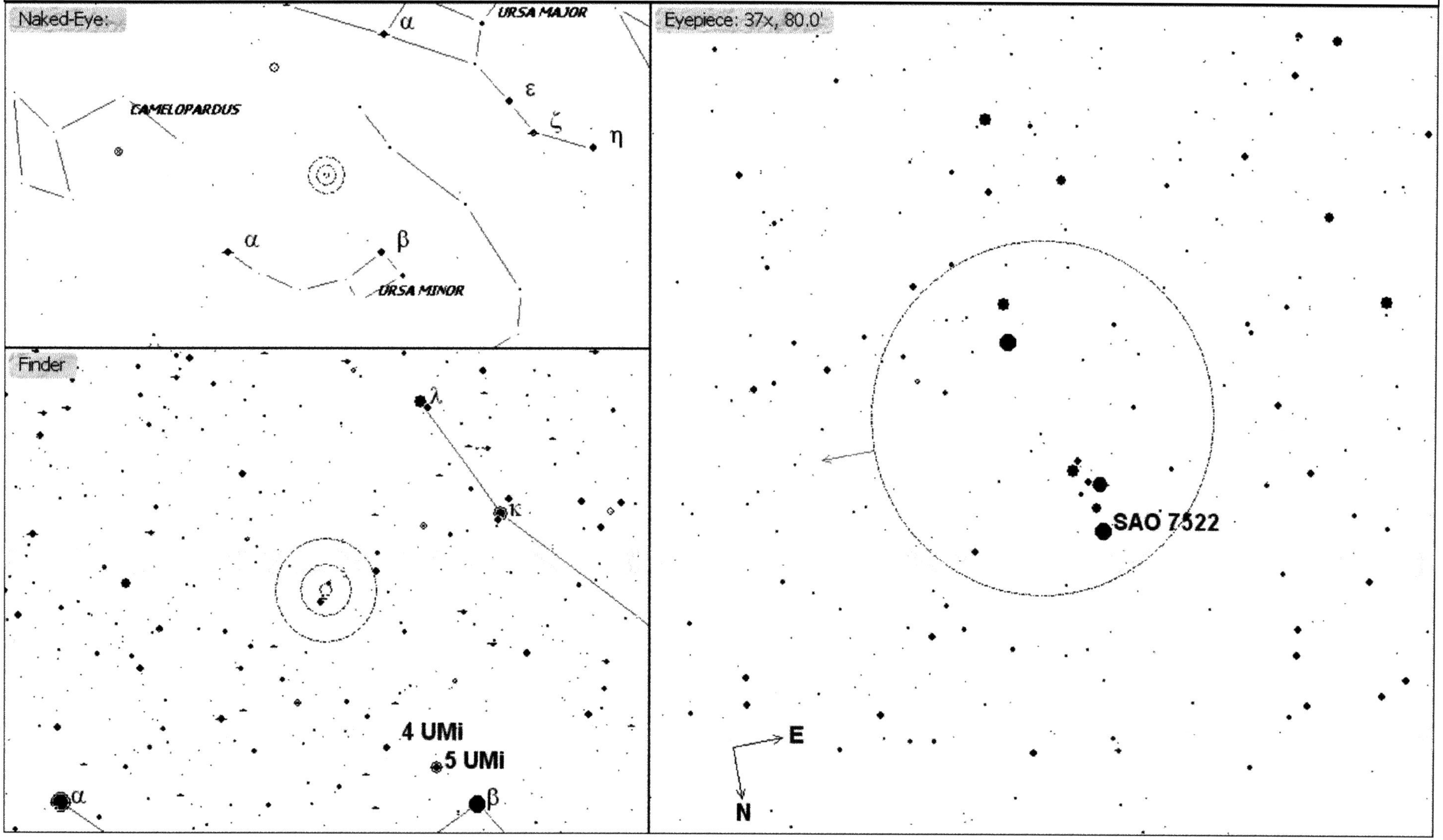

Naked-Eye:

Finder

Eyepiece: 37x, 80.0'

SAO 7522

E
N

6" f/8 Reflector

Corvus and Draco

Mid-March, 10pm. 37 degrees Latitude

Notice how the head of Draco in the northeast and Corvus in the southeast are both just above the eastern horizon. Notice the resemblance of the quadrilateral shapes. The head of Draco rises soon after Corvus.

Eight hours later, Corvus is setting, but the head of Draco has not even crossed the meridian!

Corvus lies in the 12th hour of Right Ascension, The head of Draco in the 17th.

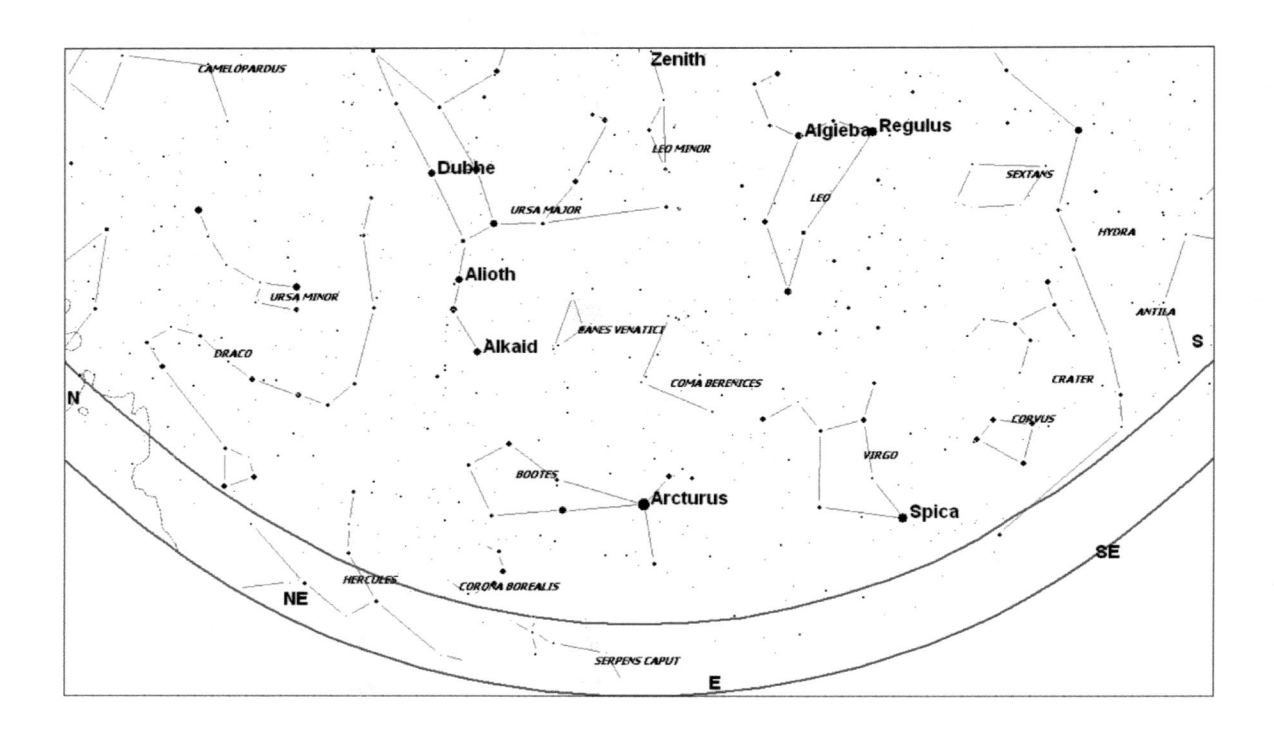

Name	**Kemble's Cascade: Chart to 9th Magnitude**
Location	RA: 03h 57m 25s, Dec: +63° 04' 20" in Camelopardalis
Found With	Binoculars
Description	This stream of stars is easily seen with binoculars. Near its southeastern end is bright open cluster NGC 1502. The distance between SAO 12925 and SAO 13012 is just nearly 2.5 degrees.

Principal Stars

SAO 12925 Single Star
aka HD 23649, PPM 14515, BD +63 00451
Magnitude: 7.92, M0

SAO 12941 Single Star
aka HD 23871, PPM 14535, BD +63 00457
Magnitude: 8.92, B8

SAO 12946 Variable Star
aka HD 23982, PPM 14542, HIP 18172,
BD +63 00458, NSV 1386
Suspected Variable, Type: UNK
Mag: V 7.94 to V 8.06, B5
Distance: 1700 +/- 830.0 ly

SAO 12951 Single Star
aka HD 24065, PPM 14546, BD +62 00618
Magnitude: 7.90, K5

SAO 12952 Single Star
aka HD 24116, PPM 14552, BD +62 00619
Magnitude: 8.93, B9

SAO 12962 Single Star
aka PPM 14564, HIP 18406, BD +62 00623
Magnitude: 8.26
Distance: 1100 +/- 380.0 ly

SAO 12960 Single Star
aka PPM 14561, HIP 18395, BD +62 00622
Magnitude: 8.70
Distance: 320 +/- 34.0 ly

SAO 12966 Single Star
aka PPM 14565, BD +62 00625, TYC 04067-0939 1
Magnitude: 8.74

*SAO 12969 Single Star
aka HD 24479, HR 1204, PPM 14575, HIP 18505
Magnitude: 5.04, G3
Distance: 340 +/- 23.0 ly

SAO 12972 Single Star
aka HD 24514, PPM 14577, HIP 18525
Magnitude: 7.98, F2
Distance: 320 +/- 32.0 ly

Object data from SKYTOOLS™ 2 — used with permission.

Name	**Kemble's Cascade to 7th Mag**
Location	RA: 03h 57m 25s, Dec: +63° 04' 20" in Camelopardalis
Found With	Binoculars
Description	This stream of stars is easily seen in binoculars. Near its southeastern end is bright open cluster NGC 1502. The distance between SAO12925 and SAO 13012 is just nearly 2.5 degrees. This view labels stars to 7th magnitude. The carbon star UV Cam is just south of SAO 13012.

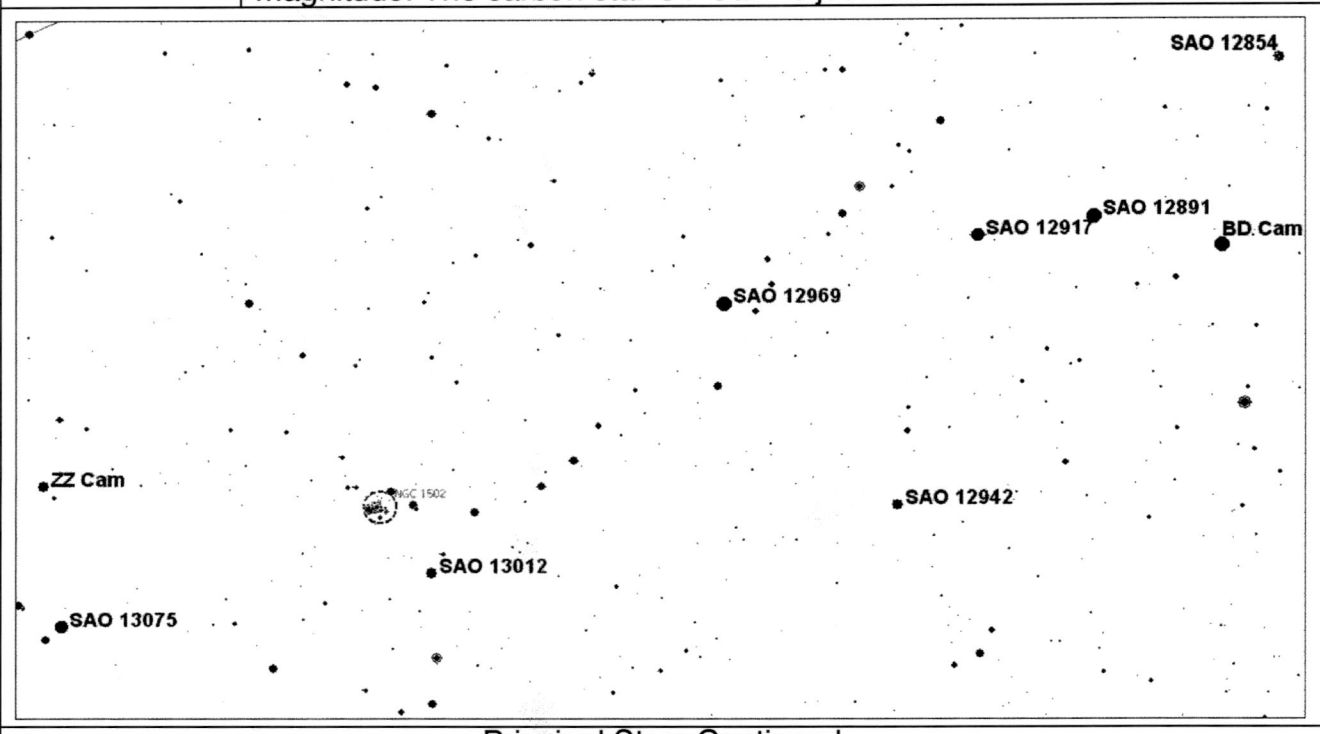

Principal Stars Continued

SAO 12981 Single Star
aka PPM 14593, BD +62 00635, TYC 04068-1253 1
Magnitude: 8.97

SAO 12987 Single Star
aka HD 24895, PPM 14604, BD +62 00639
Magnitude: 8.47, K0

HD 24992 Multiple Star System
aka HIP 18818, BD +62 00640, ADS 2924, BRD 1
Distance: 4200 +/- 11000.0 ly
AB: 8+9.51 mag,
B=HIP 18818; In 1991: PA 224° Sep 1.99"
AC: 8+10.92 mag
C=HIP 18818; In 1991: PA 174° Sep 8.77"

SAO 12998 Multiple Star System
aka HD 25090, PPM 14618, HIP 18884
Mag: H 7.40 to H 7.43, B0.5 III
Distance: 1500 +/- 850.0 ly
AB: 7.34+9.97 mag; In 1991: PA 177° Sep 0.39"

SAO 13004 Single Star
aka HD 25290, PPM 14628, BD +61 00665
Magnitude: 7.69, F5

SAO 13012 Single Star
aka HD 25443, PPM 14640, HIP 19139
Magnitude: 6.78, B2
Distance: 2200 +/- 1200.0 ly

NGC 1502 Open Cluster
aka Collinder 45, OCL 383
Magnitude: 4.10
Size: 7.0'
Distance: 3100 ly
Age: 10 Myrs
Color Excess E(B-V): 0.770

Struve 485 Multiple Star System
In Open Cluster NGC 1502
aka HR 1260, HD 25638, SAO 13031, PPM 14662
Magnitude: 6.95, B0
AB: 6.95+7.09 mag, B=SZ Cam
In 1991: PA 305° Sep 17.91"
AC: 6.95+12.9 mag, In 1902: PA 257° Sep 6.00"
AD: 6.95+10.4 mag, D=SAO 13034

UV Cam Variable Star
aka GCVS 3557, HD 25408, SAO 13009
Variability: Type: SRB
Mag: V 7.50 to V 8.10, R8
Period: 294.000000 days
Mag: H 7.60 to H 7.70
Distance: 19000 +/- 110000.0 ly

Object data from SKYTOOLS™ 2 — used with permission.

Kemble's Cascade / Camelopardalis / SAO 12969 / RA: 03h 57m 25s, Dec: +63° 04' 20" (2000)

Naked-Eye:

Eyepiece: 7x, 7.1°

NGC 1502

OCL 385

AURIGA
Capella
CAMELOPARDUS
PERSEUS
CASSIOPEIA
CEPHEUS

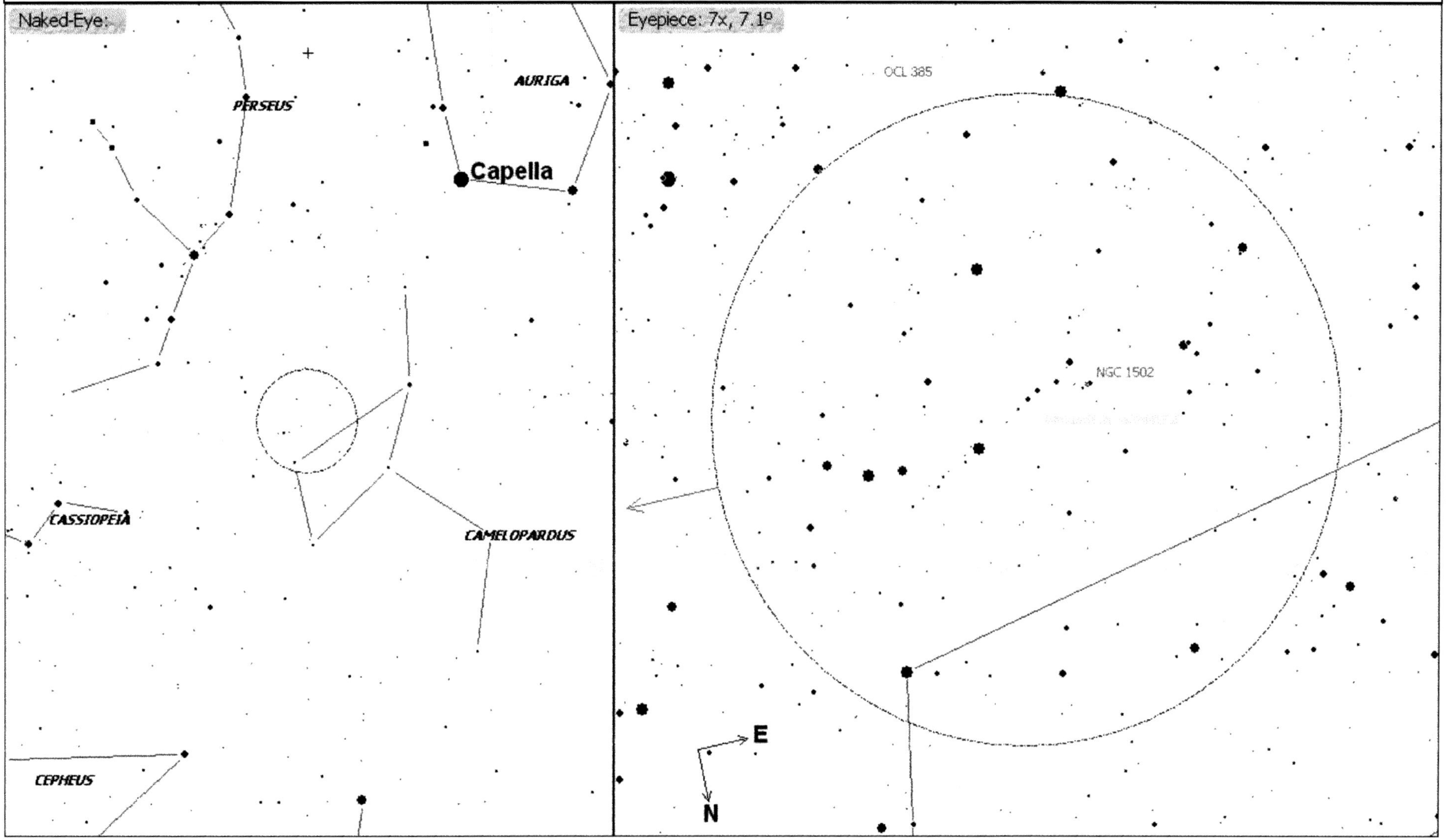

7 x 50 Binoculars

41

Name	**The 1-2 Cas Asterism**
Location	RA: 23h 06m 37s, Dec: +59° 25' 11" (2000) in Cassiopeia
Found With	Direct Observation
Description	To the eye 1 and 2 Cas appear as a fuzzy spot in the Milky Way between Cassiopeia and Cepheus. The distance between the two stars is 24 arc minutes. Optical aid reveals SAO 35152 in the field nearby. 2 Cas is one of the very multiple stars.

Principal Stars

*1 Cas   Single Star
aka HR 8797, HD 218376, SAO 35147
Magnitude: 4.85, B0.5 III
Distance: 1100 +/- 200.0 ly

SAO 35152 Single Star
aka HD 218440, HR 8803 , PPM 41498
Magnitude: 6.41, B2
Distance: 1300 +/- 310.0 ly

2 Cas   Multiple Star System
aka HR 8822, HD 218753, SAO 35186, ADS 16556, S 823
Magnitude: 5.68, A5III
Distance: 2000 +/- 670.0 ly

Component	Magnitudes	Date	Position Angle	Separation	Other Name
AB	5.68+13	1904	322°	19.9"	
AC	5.68+8.4	1824	163°	167.2"	C=HD218780
AD	5.68+8.9	1912	297°	162.7"	S 823
AP	5.68+17.7	1934	41°	10.6"	KUI 115
AQ	5.68+15	1934	278°	28.9"	KUI 115

Object data from SKYTOOLS™ 2 — used with permission.

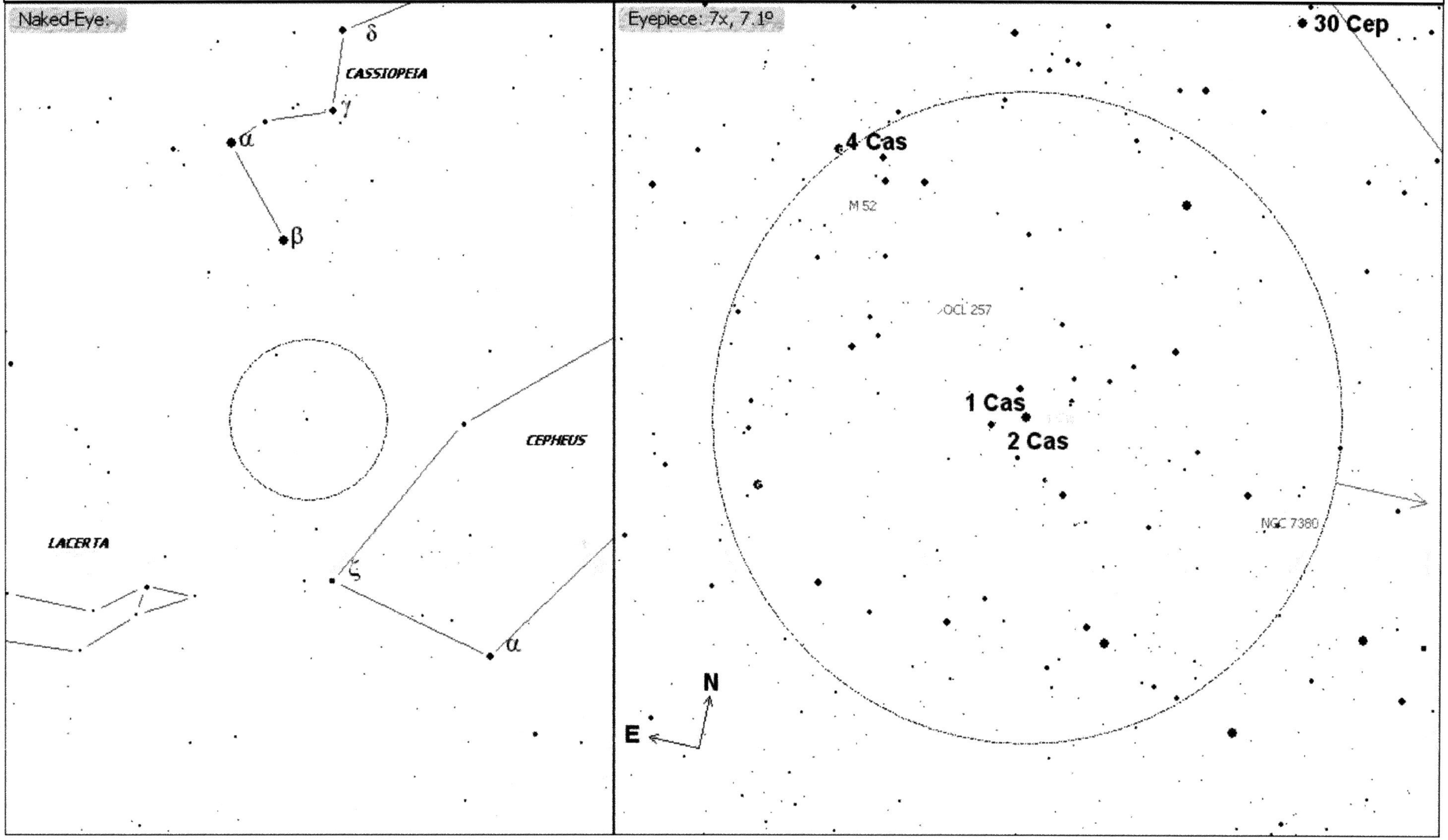

The 1-2 Cas Asterism / Cassiopeia / 1 Cas / RA: 23h 06m 37s, Dec: +59° 25' 11"

Naked-Eye:

Eyepiece: 7×, 7.1°

7 × 50 binocular

Name	**The 52-53 Cas Asterism**
Location	RA: 02h 02m 53s, Dec: +64° 54' 05" (2000) in Cassiopeia
Found With	Direct Vision
Description	The triangle of 52, 53 Cas and SAO 12076 are seen as a single fuzzy spot about a degree northwest of Epsilon Cas. The stars 53 Cas and SAO 12110 overlie the obscure open cluster Stock 5, which is a telescopic object, and is not to be confused with this asterism. SAO 12076 is brightest star here, but oddly has no Flamsteed number, and is plotted as an anonymous star on most atlases.

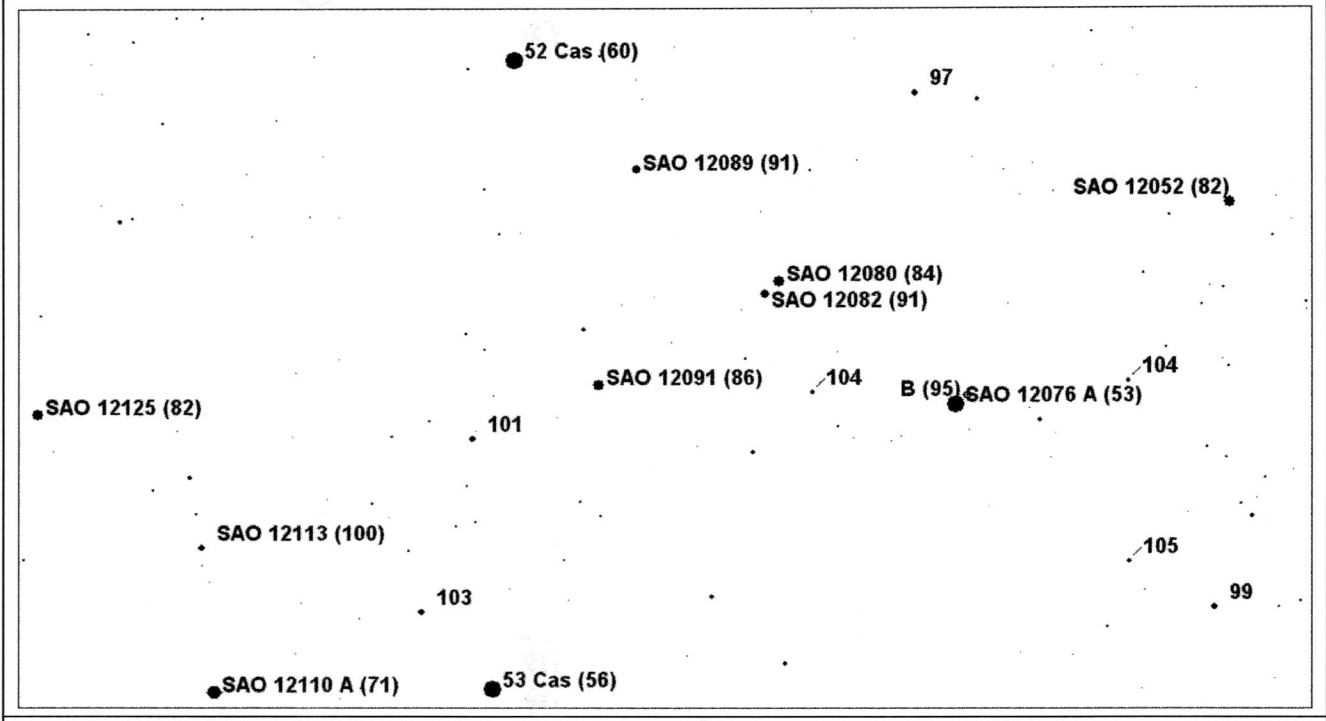

Principal Stars

SAO 12076 Multiple Star System
aka HR 567 HD 11946, ADS 157, HJ 1100
Magnitude: 5.29, A0
Distance: 260 +/- 13.0 ly
AB: 5.29+9.5 mag.
In 1959: PA 310° Sep 41.50"
SAO 12080 Single Star
aka HD 12080, PPM 13402, BD +64 00278
TYC 04041-0790 1
Magnitude: 8.41, A0
SAO 12082 Single Star
aka PPM 13403, BD +64 00279,
TYC 04041-0342-1
Magnitude: 9.07
SAO 12089 Single Star
aka HD 12183, PPM 13410, BD +64 00281
TYC 04041-0231 1
Magnitude: 9.13, A2

SAO 12091 Single Star
aka HD 12217, PPM 13416, BD +63 00273
TYC 04041-1392 1
Magnitude: 8.61, F0
52 Cas Single Star
aka HR 586, HD 12279, SAO 12095,
HIP 9564, BD +64 00282, GC 2446
Magnitude: 6.00, A1
*53 Cas Single Star
aka HR 589, HD 12301, SAO 12097, PPM 13423, HIP 9573, BD +63 00274,
TYC 04041-1956 1
Magnitude: 5.59, G6

Object data from SKYTOOLS™ 2 — used with permission.

The 52-53 Cas Asterism / Cassiopeia / 52 Cas / RA: 02h 02m 53s, Dec: +64° 54' 05"

Eyepiece: 37x, 80.0'

Naked-Eye

Finder

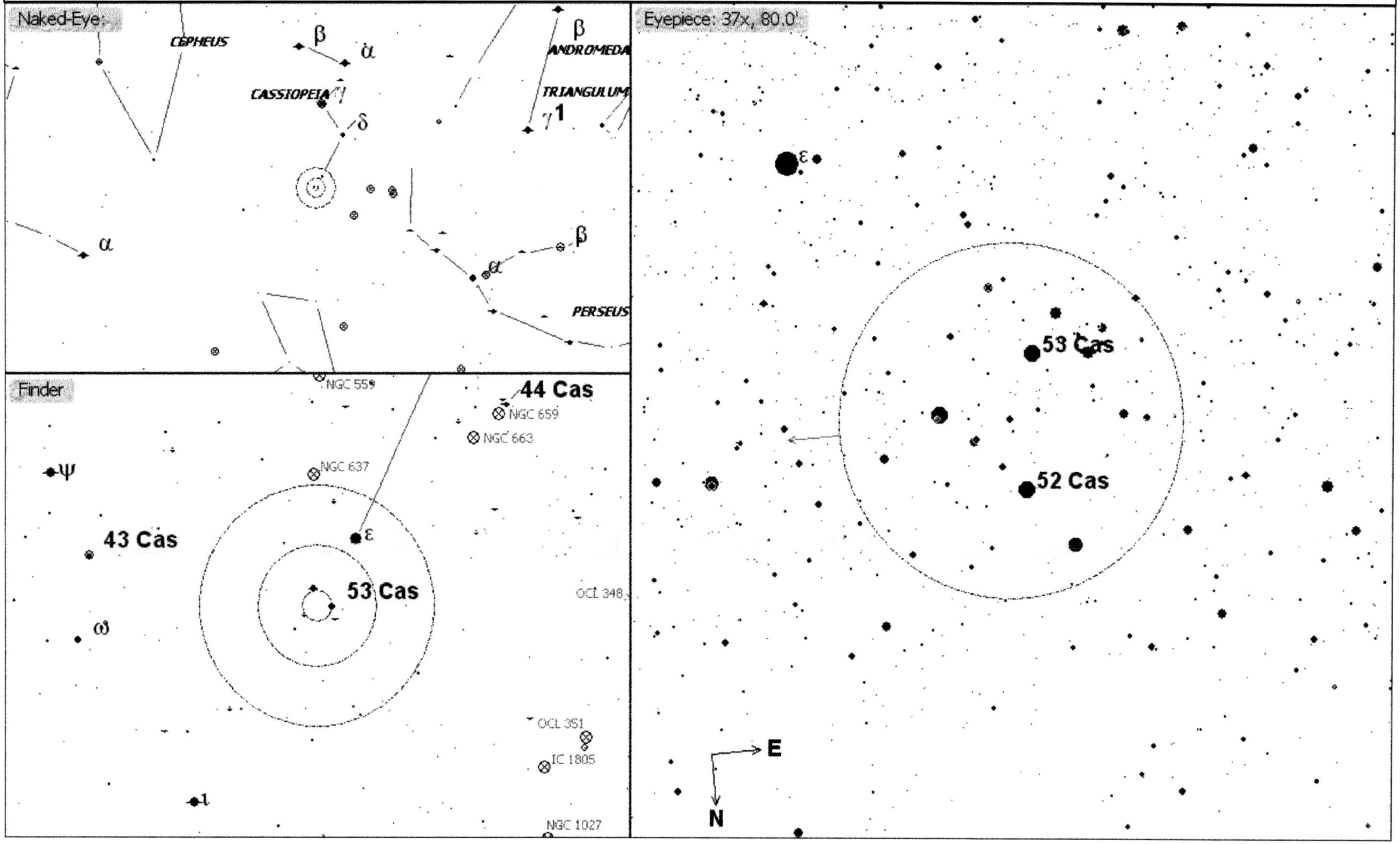

6" f/8 Reflector

45

Name	**The Rho (ρ) Cass. Asterism**
Location	RA: 23h 54m 23s, Dec: +57° 29' 58" (2000) in Cassiopeia
Found With	Rho Cas
Colorful Stars	Direct Vision
Description	The pair of Rho Cass and V373 Cas appear as a fuzzy spot in the sky. The distance between them is 11'. Rho has a lovely yellow color and undergoes changes in brightness and spectral type. Bright open cluster NGC 7789 is just a degree to the southeast.

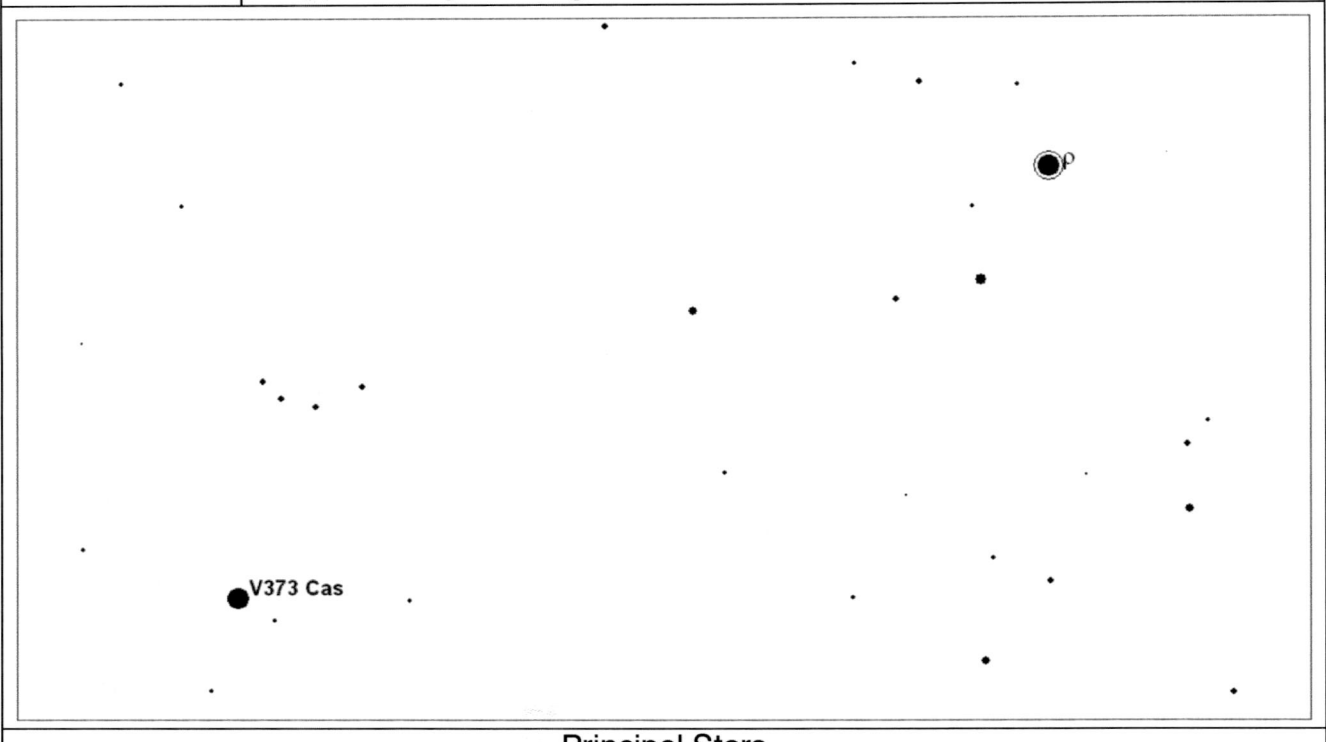

Principal Stars

***Rho Cas**   Variable Star
aka 7 Cas, GCVS 5256, HR 9045, HD 224014, SAO 35879
Magnitude: 4.59 G2
Variability: Mag V 4.10 to V 6.20,
Period: 320.000000 days

NGC 7789   Open Cluster
aka OCL 269,
Magnitude: 7.50,   Size: 15.0'
Distance: 6200 ly
Age: 1770 Myrs

V373 Cas   Variable Star
aka GCVS 4983, HR 9052, HD 224151 SAO 35899
Magnitude: 6.03, B0.5 II
Variability: Mag V 5.90 to V 6.30
Period: 13.419200 days
Epoch: JD 2436491.2370

Object data from SKYTOOLS™ 2 — used with permission.

The Rho (ρ) Cas Asterism / Cassiopeia / RA: 23h 54m 23s, Dec: +57° 29' 58"

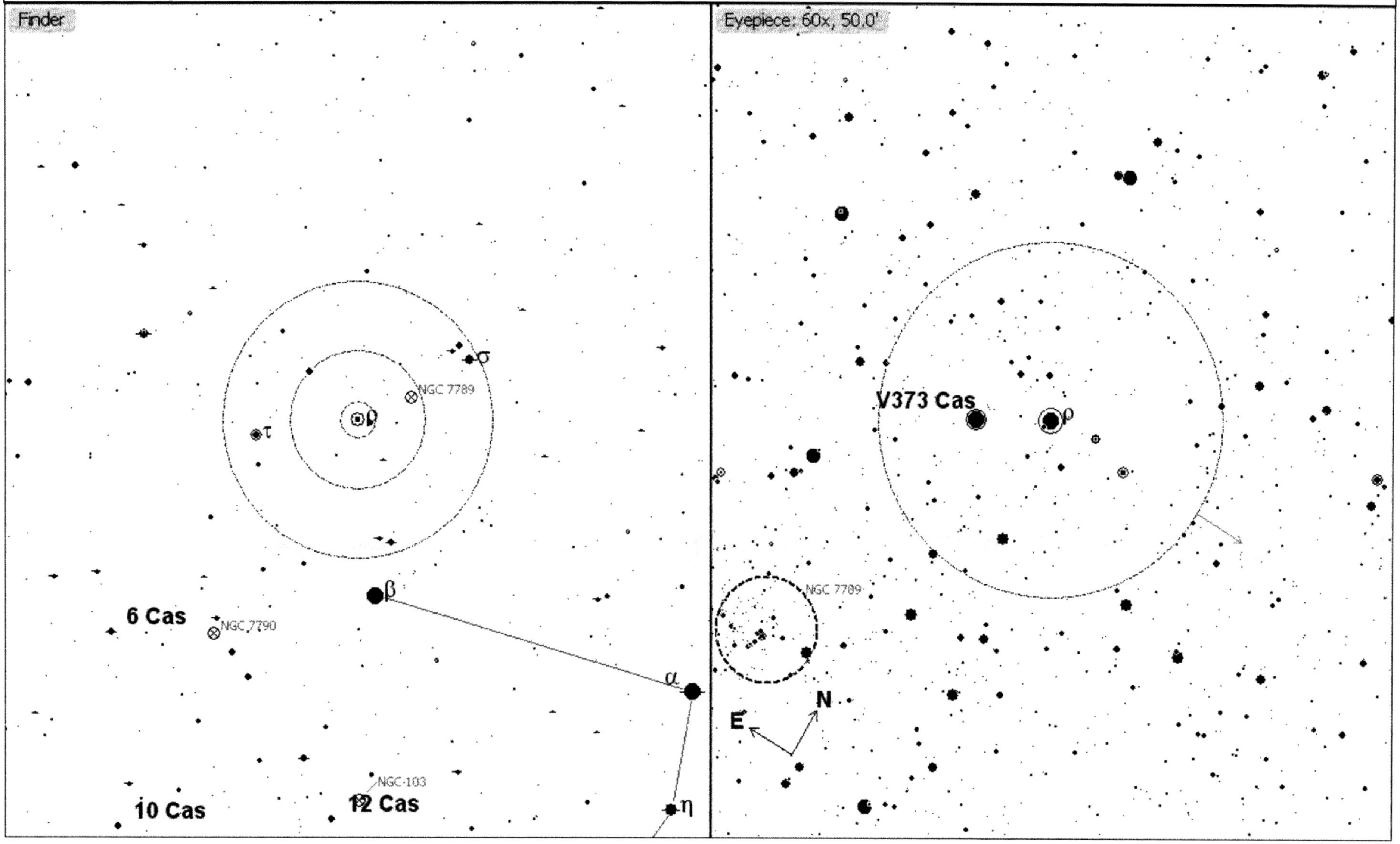

Finder

Eyepiece: 60x, 50.0'

6" f/8 Reflector

Name	**The Sigma (σ) Cas Asterism**
Location	RA: 23h 59m 00s, Dec: +55° 45' 18" (2000) in Cassiopeia
Found With	SAO 35913
Colorful Stars	Direct Vision
Description	Sigma Cas and SAO 35917 are seen as a fuzzy spot in dark skies. With just the slightest optical aid SAO 35913 is seen as a third member of this asterism. The distance from Sigma to SAO 35917 is 16'. The wonderful open cluster NGC 7789 is nearby, between Sigma and Rho Cas.

Principal Stars

SAO 35913   Multiple Star System
aka HD 224320, PPM 42377, HIP 118057,
BU 1224, ADS 17114
Magnitude: 6.56, K0
Distance: 2500 +/- 1400.0 ly
AB: 6.56+13.7 mag
In 1890: PA 203° Sep 4.00"
AC: 6.56+12.1 mag
In 1912: PA 357° Sep 77.20"

NGC 7789   Open Cluster
aka OCL 269
Magnitude: 7.50, Size: 15.0'
Distance: 6200 ly
Age: 1770 Myrs

SAO 35917   Variable Star
aka HR 9059, HD 224355, PPM 42381
Magnitude: 5.57, G8I
Distance: 220 +/- 10.0 ly

*Sigma Cas   Multiple Star System
aka 8 Cas, HR 9071, HD 224572,
HIP 118243, Struve 3049, ADS 17140
Magnitude: 4.88, B1
Distance: 1500 +/- 530.0 ly
AB: 4.88+7.33 mag, B=HD 224572;
In 1991: PA 326° Sep 3.14"

Object data from SKYTOOLS™ 2 — used with permission.

The Sigma (σ) Cas Asterism / Cassiopeia / RA: 23h 59m 00s, Dec: +55° 45' 18"

Finder

Eyepiece: 60x, 50.0

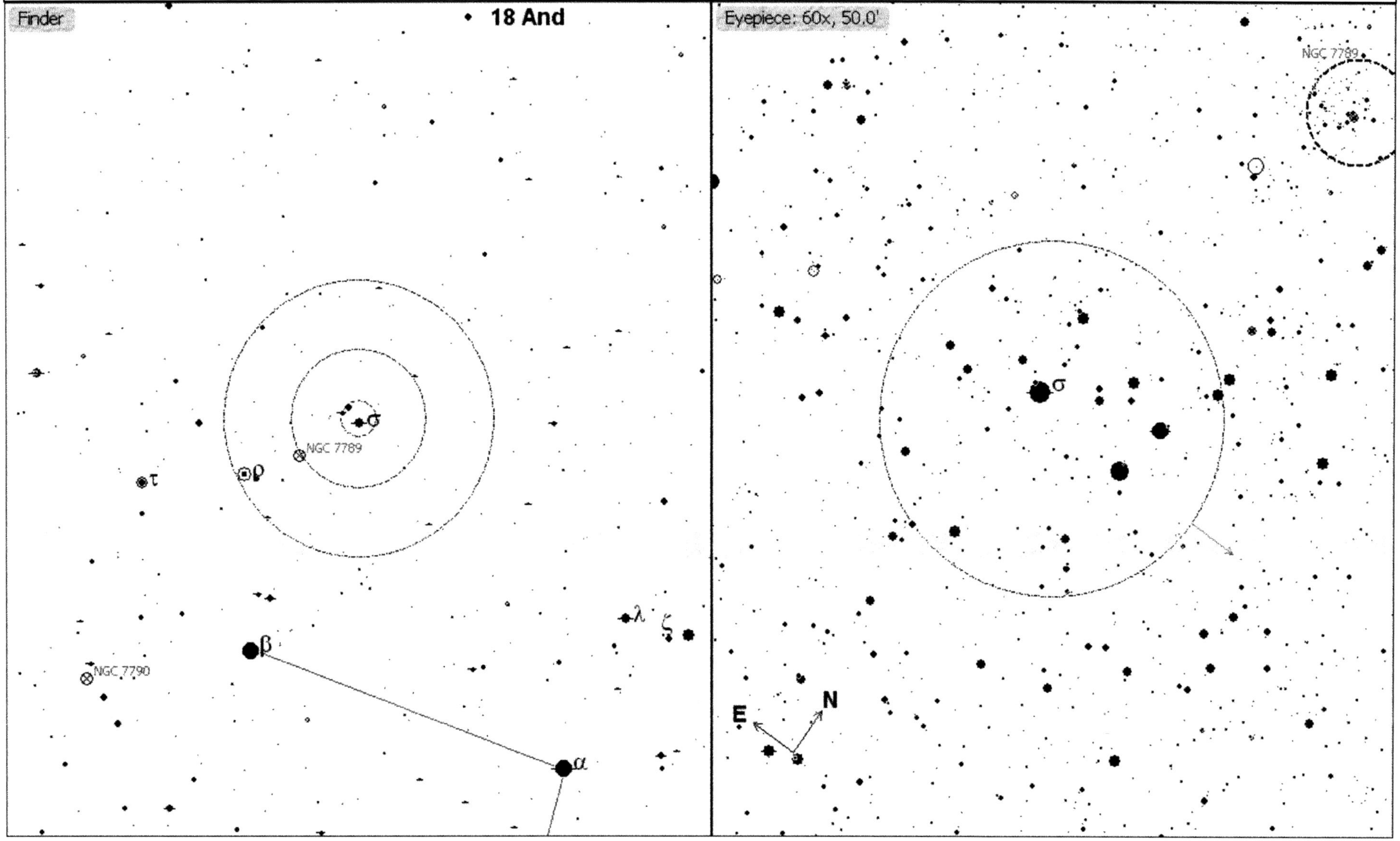

6" f/8 Reflector

Name	**The Theta - Mu (θ - μ) Cass. Asterism**
Location	RA: 01h 11m 06s, Dec: +55° 09' 00" (2000) in Cassiopeia
Found With	Mu Cas, SAO 22044
Colorful Stars	Direct Vision
Description	The Theta-Mu pair appears as a fuzzy spot to the eye in a dark sky. Closer inspection reveals two fainter stars in the vicinity, forming an attractive pyramid shaped asterism. The distance between Theta and Mu is 28', almost half of a degree.

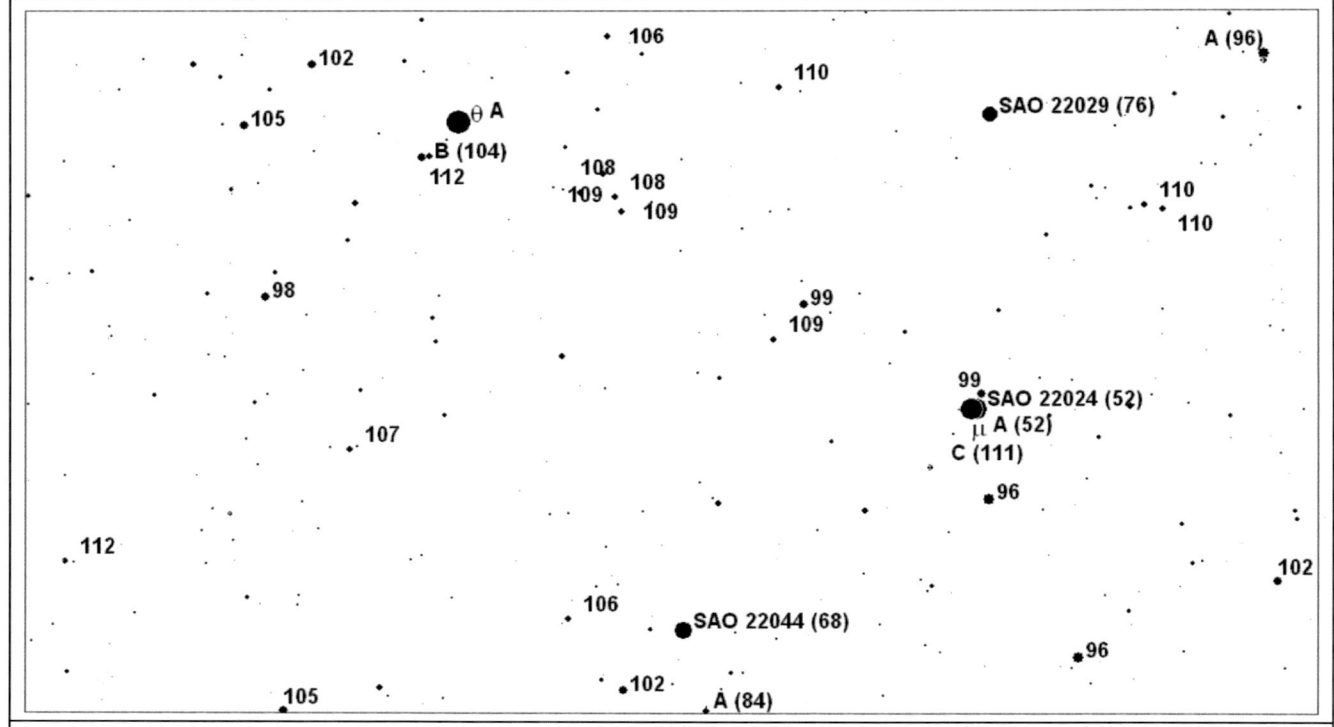

Principal Stars

Mu μ Cas Multiple Star System
aka, 30 Cas, HR 321, HD 6582, SAO 22024
Magnitude: 5.18, G5
AB: 5.3+11.1 mag
Orbit: P=21.4 yr, a=0.19"
PA 243° Sep 0.23" (2004.8)
AC: 5.3+11.1 mag
In 1907: PA 145° Sep 205.80"
CD: 11.1+12.8 mag
In 1907: PA 114° Sep 4.20"
AP: 5.3+12.3 mag
In 1907: PA 145° Sep 87.70"

SAO 22029 Single Star
aka HD 6656, PPM 26044, HIP 5333
Magnitude: 7.61, F2
Distance: 280 +/- 21.0 ly

SAO 22044 Single Star
aka HD 6833, PPM 26062, HIP 5458
Magnitude: 6.75, K0III
Distance: 750 +/- 130.0 ly

*Theta θ Cas Multiple Star System
aka, 33 Cas, HR 343, HD 6961, SAO 22070
Magnitude: 4.34, A6
Distance: 140 +/- 3.9 ly
AB: 4.34+10.4 mag
In 1907: PA 135° Sep 145.60"

Object data from SKYTOOLS™ 2 — used with permission.

The Theta - Mu (θ - μ) Cas Asterism / Cassiopeia / RA: 01h 11m 06s, Dec: +55° 09' 00"

Finder

Eyepiece: 60x, 50.0'

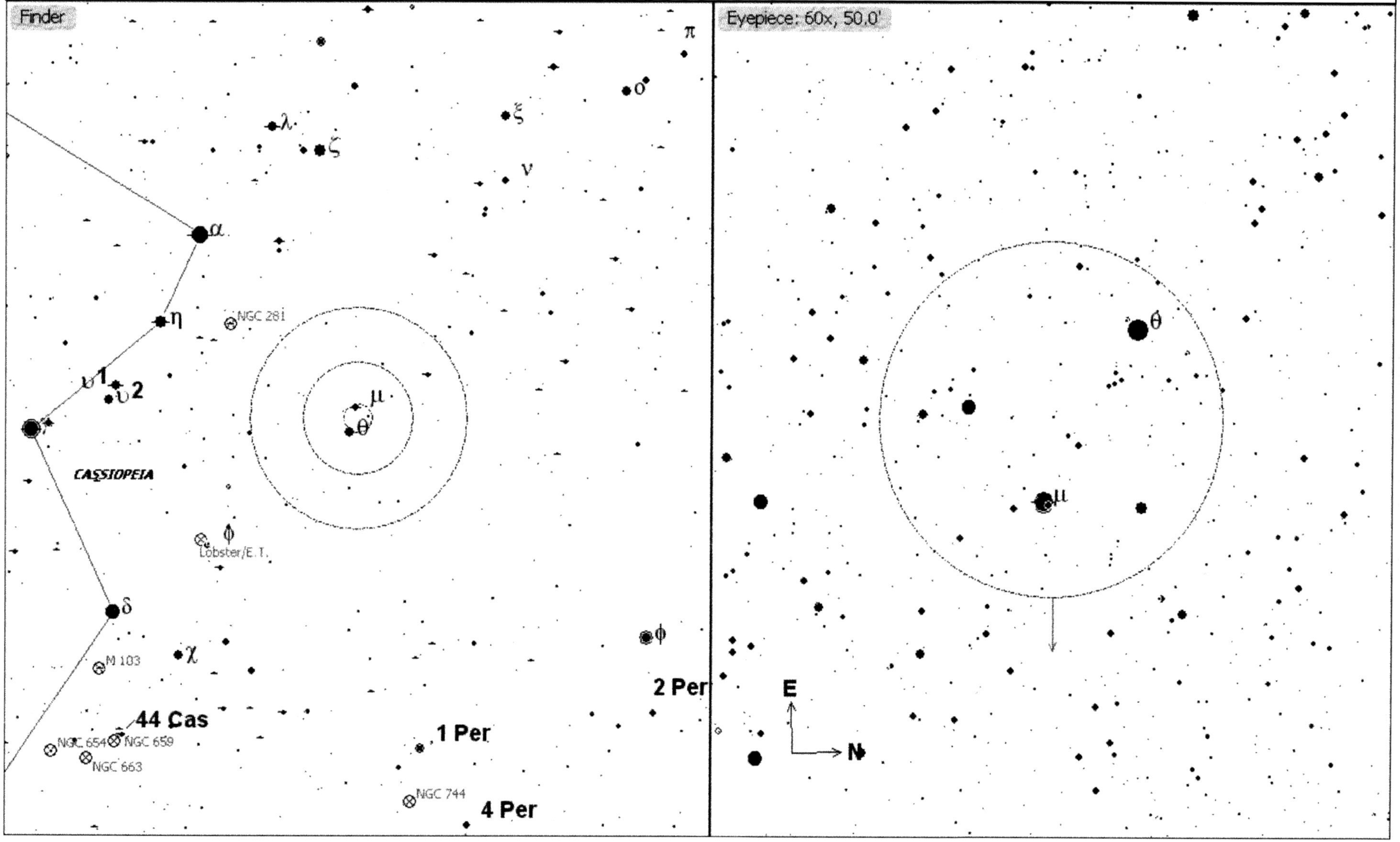

6" f/8 Reflector

Name	**The Upsilon (υ) Cas Asterism**
Location	RA: 00h 55m 00s, Dec: +58° 58' 22" (2000) in Cassiopiea
Colorful Stars	All four: Upsilon 1 & 2; SAO 21841, 21851
Found With	Direct Vision
Description	The pair of Upsilon-1 and Upsilon-2 Cas are seen as a fuzzy spot to the eye. Closer inspection reveals a pair of stars between them. The distance between Upsilon 1 and 2 is 18'. North is up.

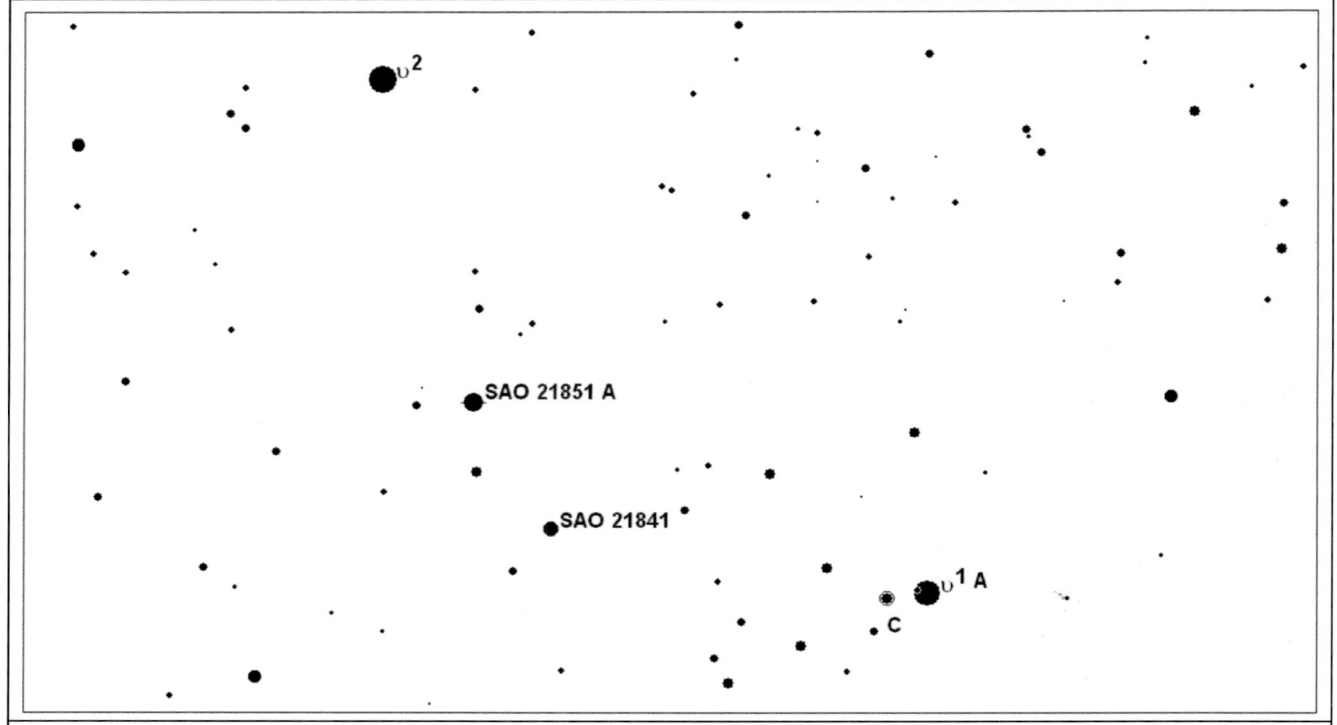

Principal Stars

*Upsilon ¹ Cas Multiple Star System
aka 26 Cas, HR 253, HD 5234, SAO 21832,
BU 1098, ADS 748
Magnitude: 4.84, K2III
Distance: 410 +/- 33.0 ly
AB: 4.83+12.5 mag
In 1934: PA 70° Sep 14.30"
AC: 4.83+11.8 mag
In 1959: PA 127° Sep 94.10"
SAO 21841 Single Star
aka HD 236577 , PPM 25842
Magnitude: 9.49, K0

SAO 21851 Multiple Star System
aka HD 236579 , PPM 25851, HIP 4397
Magnitude: 8.13, K0III
Distance: 1100 +/- 350.0 ly
AB: 8.13+10.59 mag
In 1991: PA 270° Sep 0.17"
Upsilon ² Cas Single Star
aka 28 Cas, HR 265, HD 5395, SAO 21855
Magnitude: 4.64, G8III
Distance: 210 +/- 7.5 ly

Object data from SKYTOOLS™ 2 — used with permission.

The Upsilon (υ) Cas Asterism / Cassiopeia / υ¹ Cas / RA: 00h 55m 00s, Dec: +58° 58' 22"

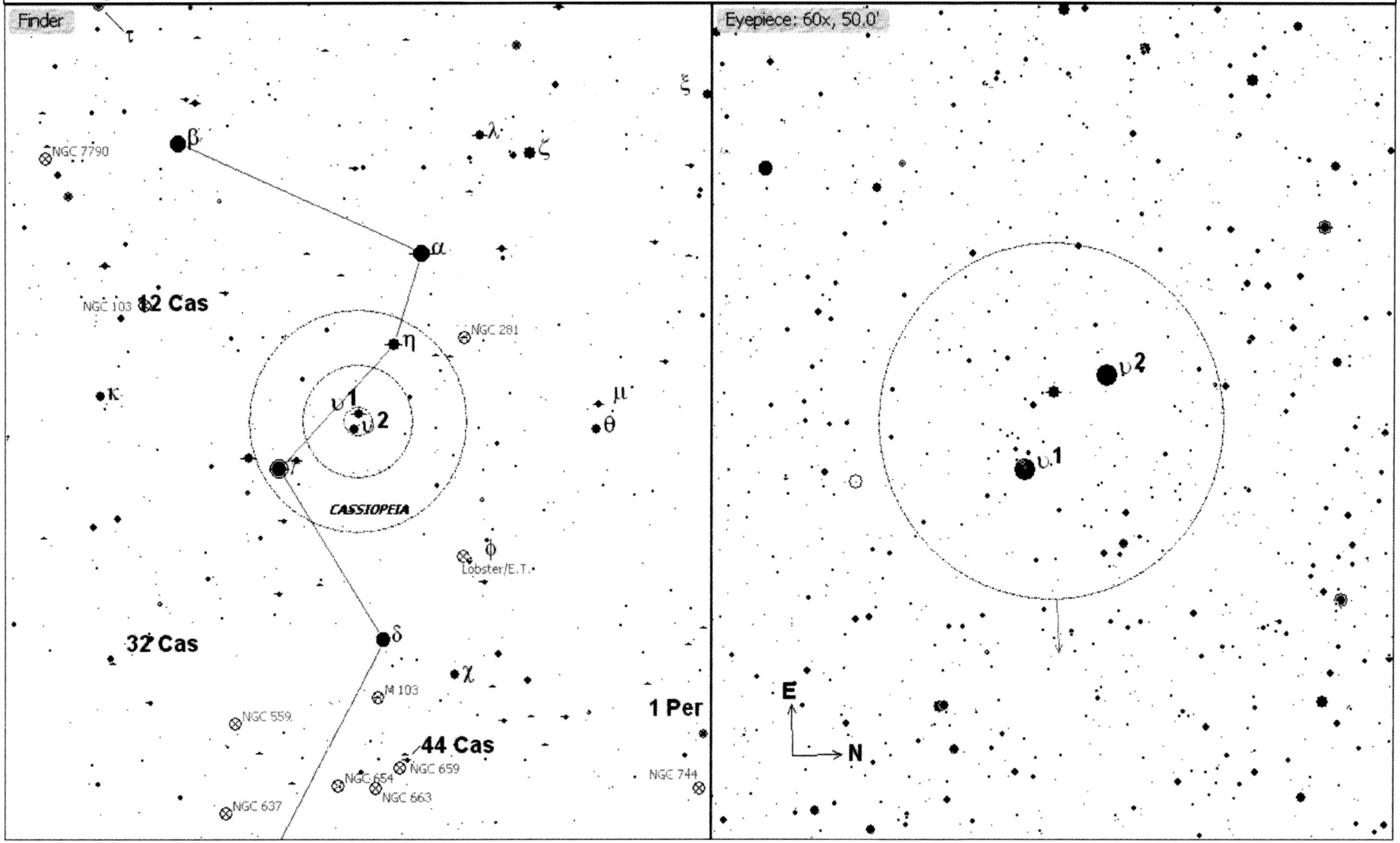

Finder

Eyepiece: 60x, 50.0'

6" f/8 Reflector

Name	**The Zeta (ζ) Cass. Asterism**
Location	RA: 00h 36m 58s, Dec: +53° 53' 49" (2000) in Cassiopeia
Found With	Direct Vision
Description	Zeta Cass and SAO 21551 form a naked eye pair 18' apart. This closeness makes the pair somewhat difficult to resolve visually. In dark skies this group appears as a fuzzy spot to the eye. Other wide pairs: Epsilon Lyrae (the Double Double) a visual pair 3'30" apart, and Alpha 1-2 Capriconii at 6'20" apart. With a small scope the two fainter SAO stars help form an attractive grouping.

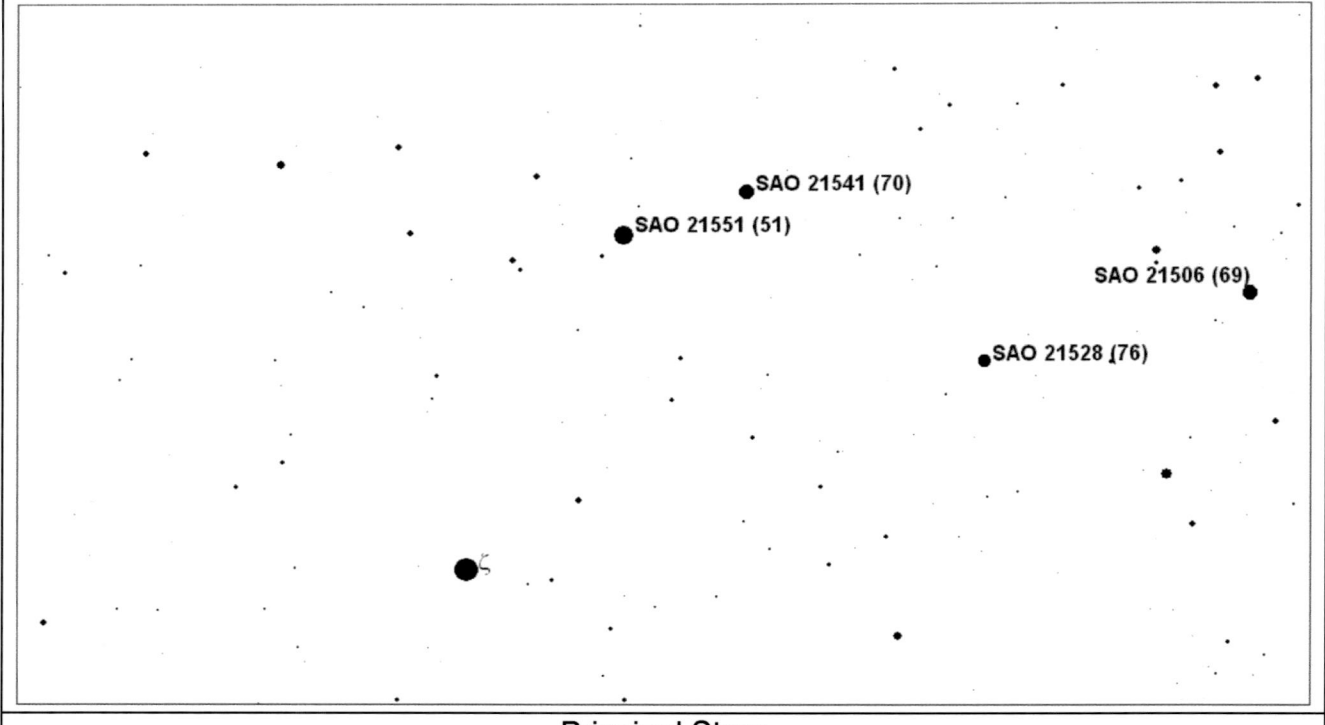

Principal Stars

SAO 21528 Single Star
aka HD 3051, PPM 25483, HIP 2697
Magnitude: 7.60, A0
Distance: 800 +/- 160.0 ly

SAO 21541 Single Star
aka HD 3184, PPM 25498, HIP 2782
Magnitude: 6.95, A2
Distance: 380 +/- 33.0 ly

SAO 21551 Single Star
aka HR 144, HD 3240, PPM 25509
Magnitude: 5.08, B7III
Distance: 510 +/- 49.0 ly

*****Zeta Cas** Variable Star
aka 17 Cas, HR 153, HD 3360, SAO 21566
Suspected Variable, Type: UNK
Mag: V 3.59 to V 3.68, B2IV
Distance: 600 +/- 67.0 ly

Object data from SKYTOOLS™ 2 — used with permission.

The Zeta (ζ) Cas Asterism / Cassiopeia / RA: 00h 36m 58s, Dec: +53° 53' 49"

Finder

Eyepiece: 60x, 50.0'

22 And

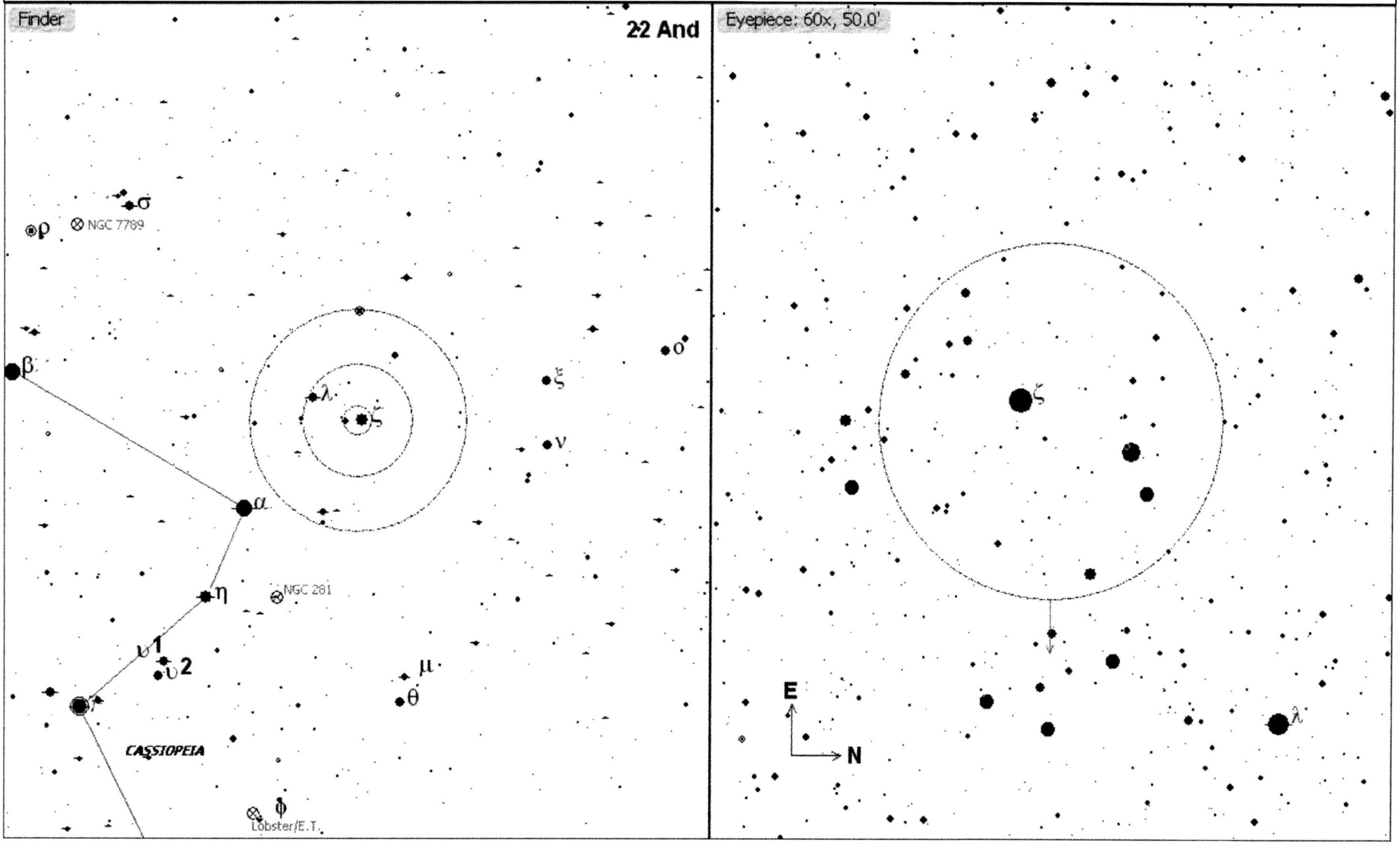

6" f/8 Reflector

Name	**15 Cephei**, Otto Struve 461, OΣ 461, SAO 34016, ADS 15601
Location	RA: 22h 03m 54s, Dec: +59° 48 '52" (2000) in Cepheus
Found With	Binoculars
Description	This wonderful multiple star appears as a bright fuzzy spot in binoculars near Zeta Cephei. South is up in this view. The separation between the A and F components is 7' 9" or 429". Compare to M73 and NGC 2017.

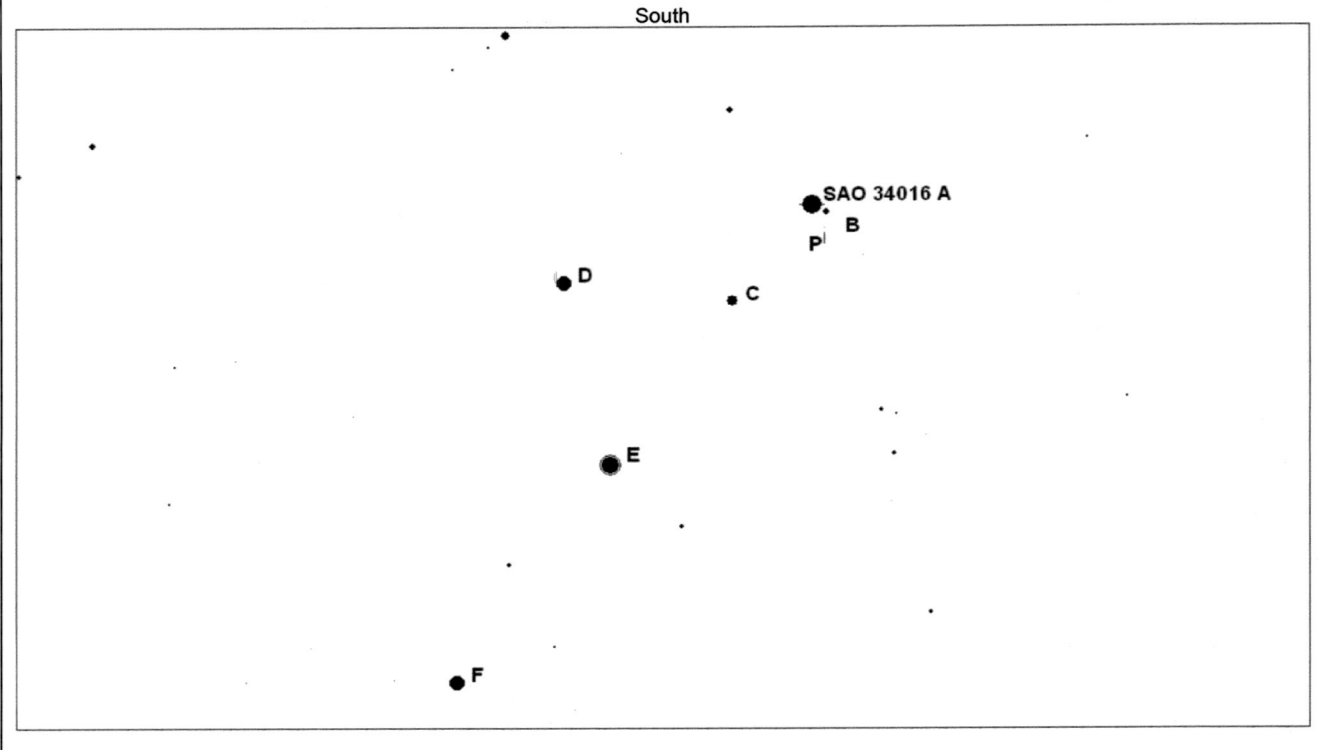

Principal Stars

Otto Struve 461 Multiple Star System
aka SAO 34016, HD 209744, PPM 40173, HIP 108925
Magnitude: 6.69, B1
Distance: 1400 +/- 430.0 ly

Component	Magnitudes	Date	Position Angle	Separation	Other Name
AB	6.7+11.4	In 1848	PA 298°	11.10"	
AC	6.7+9.7	In 1876	PA 39°	89.90"	C=SAO 34018
AD	6.7+7.9	In 1876	PA 72°	183.90"	D=HD 209810
AE	6.7+7	In 1876	PA 37°	236.70"	E=HD 209809
EF	7+8	In 1876	PA 34°	192.40"	F=HD209830
AG	6.7+9.9	In 1991	PA 230°	0.36"	
AP	6.7+14.3	In 1905	PA 334°	18.30"	

Object data from SKYTOOLS™ 2 — used with permission.

15 Cephei / SAO 34016 / RA: 22h 03m 54s, Dec: +59° 48' 52"

Naked-Eye: CASSIOPEIA, α, URSA MINOR, β, DRACO, CEPHEUS, LACERTA

Finder: 20 Cep, 18 Cep, 19 Cep, NGC 7160, 9 Cep, ν, μ, ξ, δ, ε, λ

Eyepiece: 37x, 80.0'

N, E

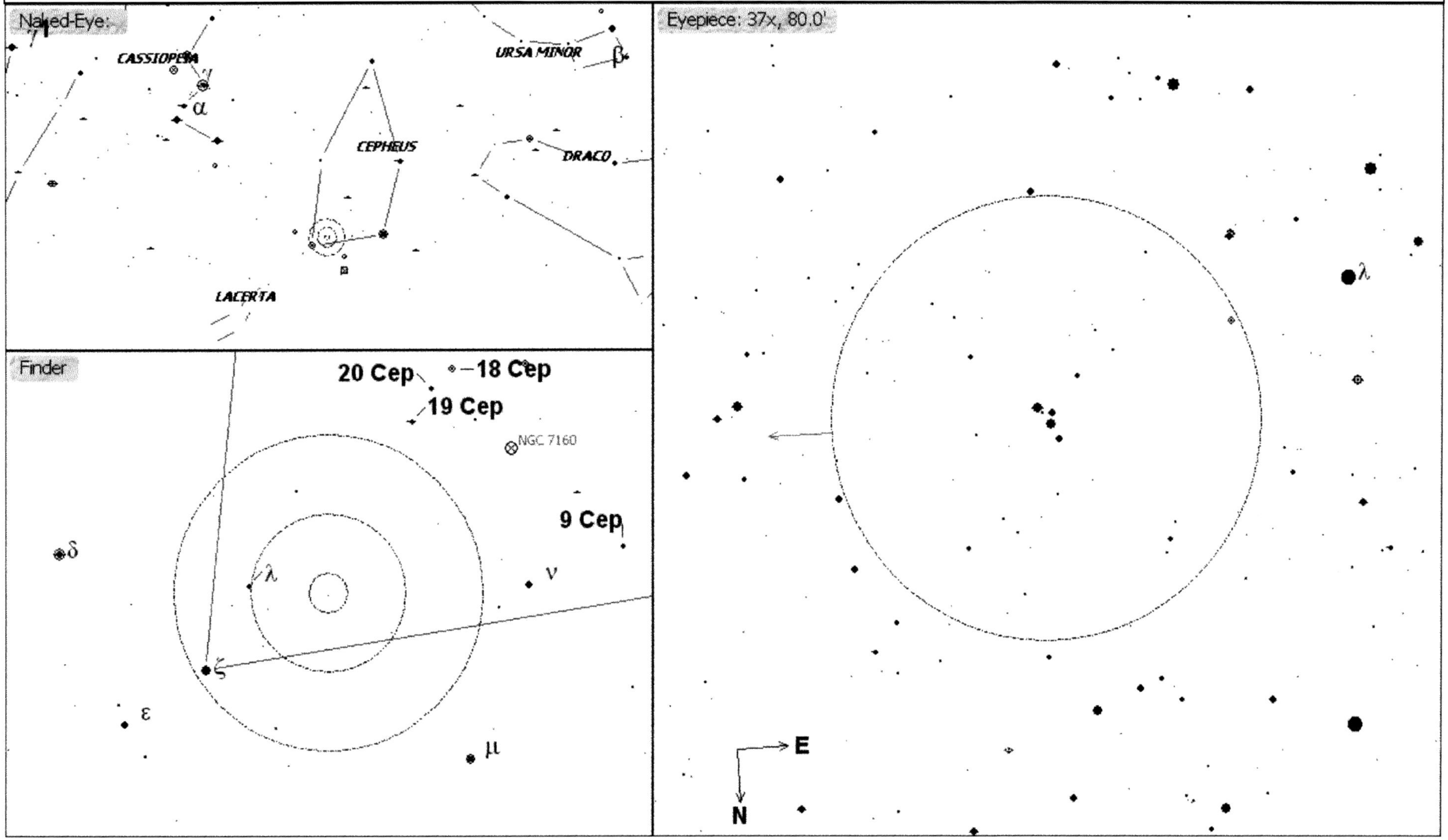

6" f/8 Reflector

57

Name	**The Eta (η) Ceti Asterism**
Location	RA: 01h 08m 35s, Dec: -10° 10' 56" (2000) in Cetus
Found With	Direct Vision
Colorful Stars	Eta, 27 Cet
Description	The brightest of the naked eye asterisms in Cetus. Binoculars or a small telescope resolve the asterism into 4 bright stars and a sprinkling of lesser lights. The distances between η and 30 Cet = 26 arc min; η to 27 Cet = 45 arc min.

Principal Stars

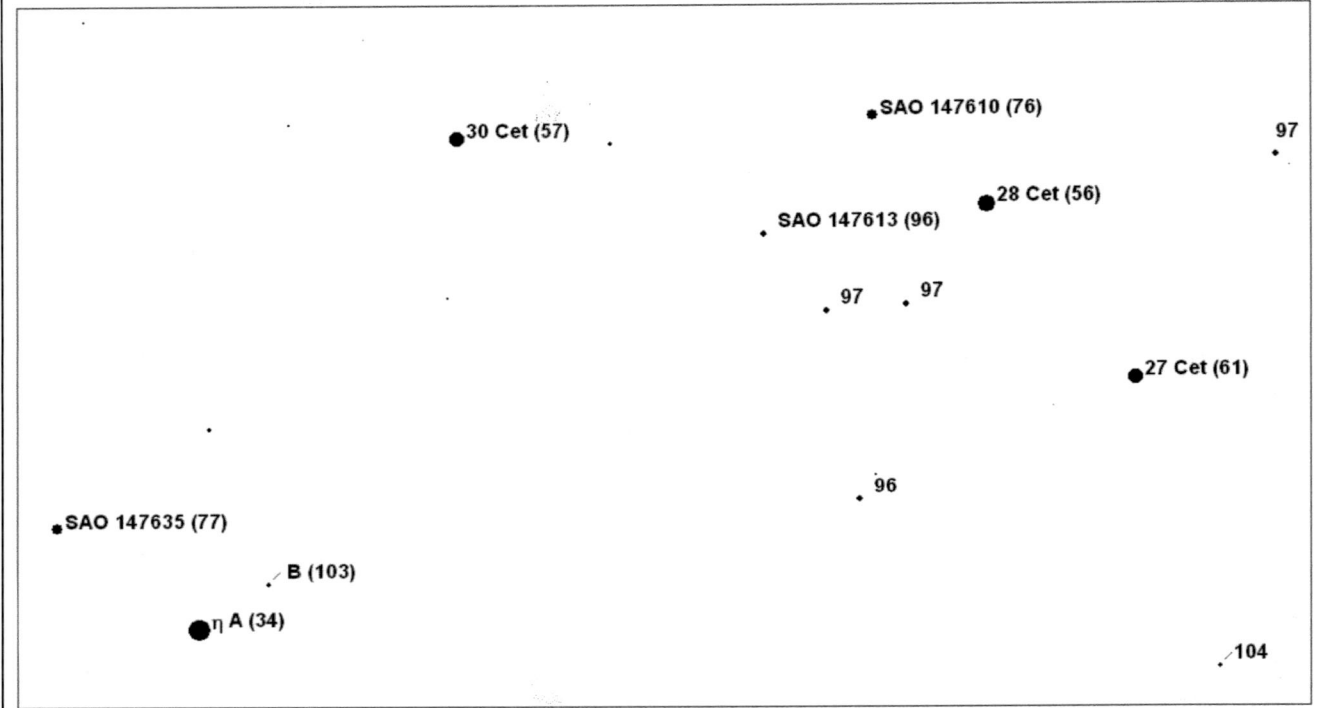

27 Cet Single Star
aka HR 315, HD 6482, SAO 147601
Magnitude: 6.12, K0III
Distance: 310 +/- 21.0 ly

28 Cet Single Star
aka HR 317, HD 6530, SAO 147606
Magnitude: 5.58, A0IV
Distance: 590 +/- 74.0 ly

SAO 147610 Single Star
aka HD 6573, PPM 183044, HIP 5197
Magnitude: 7.59, A2
Distance: 560 +/- 73.0 ly

SAO 147613 Single Star
aka PPM 183051, BD -10 00235
Magnitude: 9.61, F2

30 Cet Single Star
aka HR 329, HD 6706, SAO 147622, PPM 183061, HIP 5296,
Magnitude: 5.71, F5IV
Distance: 160 +/- 4.9 ly

*Eta η Cet Multiple Star System
aka Dheneb, 31 Cet, HR 334, HD 6805, SAO 147632, PPM 209790, HIP 5364
Magnitude: 3.44, K3III
Distance: 120 +/- 3.0 ly
AB: 3.46+10.3 mag, BUP, B=PPM 209786
In 1907: PA 305° Sep 233.5"

SAO 147635 Single Star
aka HD 6866, PPM 209796, HIP 5396
Magnitude: 7.70, K0
Distance: 870 +/- 190.0 ly

Object data from SKYTOOLS™ 2 — used with permission.

The Eta (η) Ceti Asterism / Cetus / RA: 01h 08m 35s, Dec: -10° 10' 56"

Eyepiece: 37x, 80.0'

Naked-Eye:

Finder

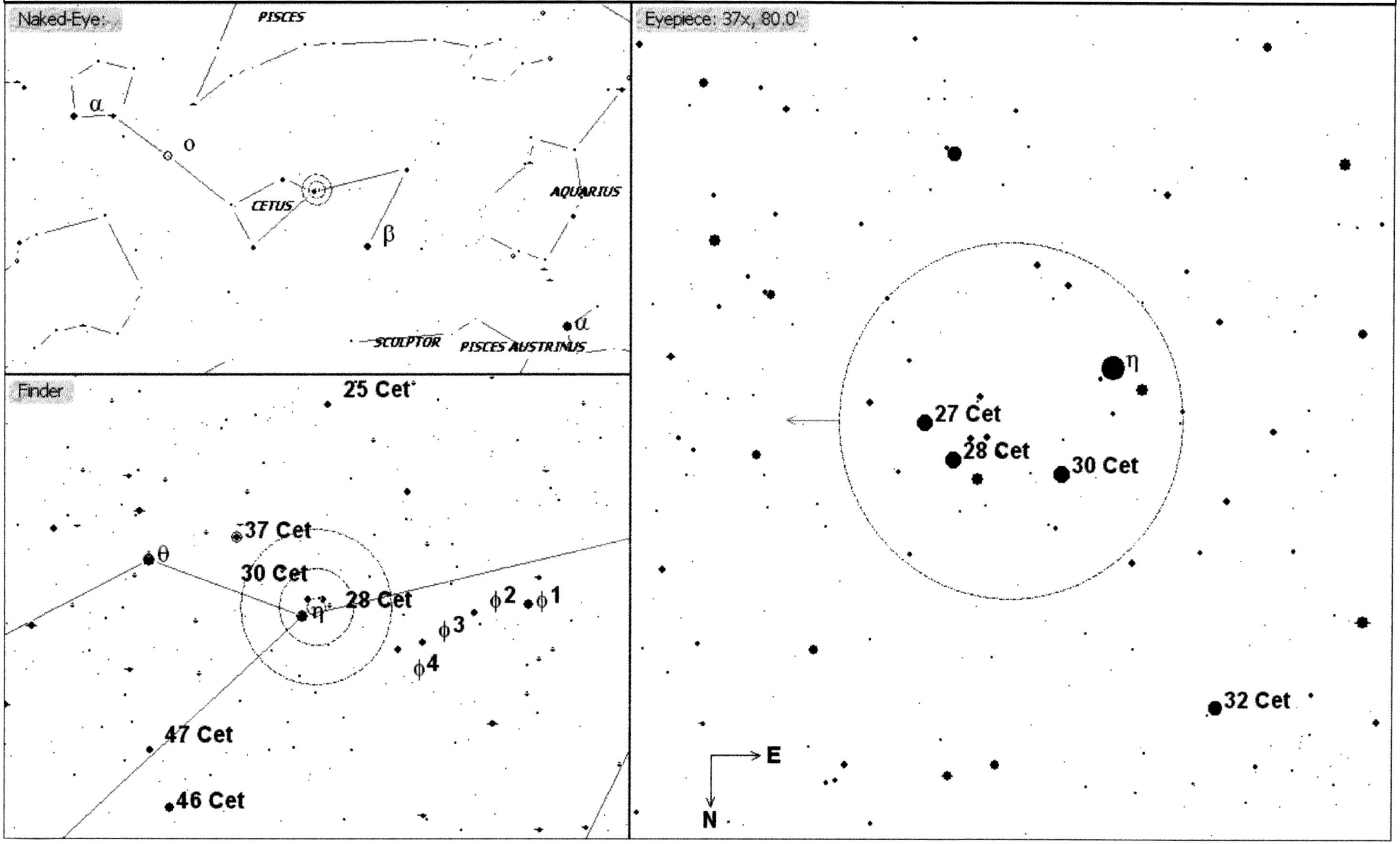

6" f/8 Reflector

Name	**The 49 – 50 Ceti Asterism**
Location	RA: 01h 35m 59s, Dec: -15° 24' 01" (2000) in Cetus
Found With	Direct Vision
Colorful Stars	50 Cet
Description	The pair of 49 and 50 Cet forms a fuzzy appearing object seen in dark skies. The distance between the two stars is about 25 arc-minutes. With slightest optical aid the difference in colors is readily seen.

Principal Stars	
49 Cet Single Star aka HR 451, HD 9672, SAO 147886, PPM 210365, HIP 7345, BD -16 00265 Magnitude: 5.62, A1 Distance: 200 +/- 9.2 ly	***50 Cet** Single Star. aka HR 459, HD 9856, SAO 147901 PPM 210401, HIP 7450, BD -16 00270 Magnitude: 5.41, K1III Distance: 550 +/- 67.0 ly

Object data from SKYTOOLS™ 2 — used with permission.

The 49 – 50 Ceti Asterism / Cetus / 50 Cet / RA: 01h 35m 59s, Dec: -15° 24' 01" (2000)

Eyepiece: 24x, 2.1°

Naked-Eye

Finder

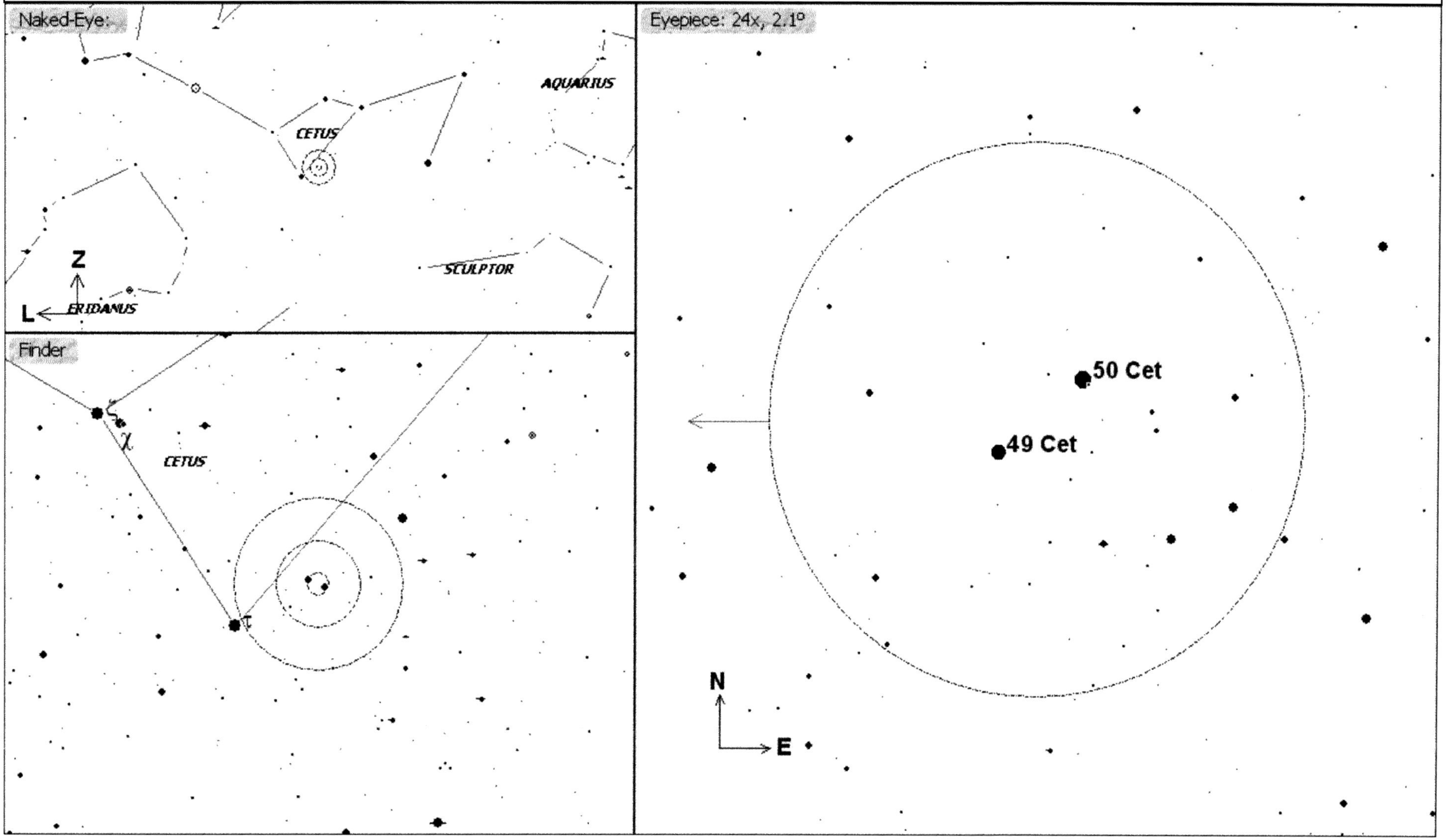

80mm f/6 Refractor

Name	**The Zeta – Chi (ζ - χ) Ceti Asterism**
Location	RA: 01h 51m 28s, Dec: -10° 20' 06" (2000) in Cetus
Found With	Direct Vision
Colorful Stars	Zeta Cet
Description	This pair of bright stars appears as a fuzzy spot in Cetus. The distance between Chi and Zeta is 34 arc minutes.

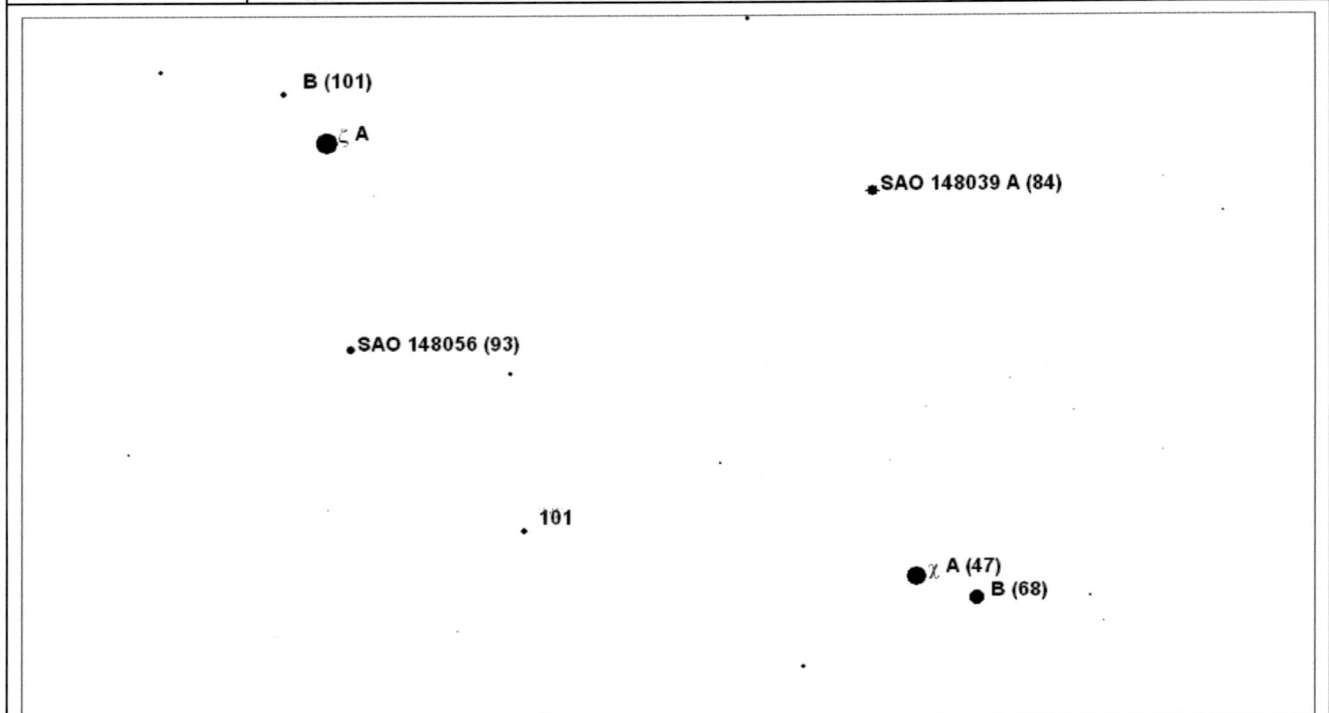

Principal Stars	
*Zeta (ζ) Cet Multiple Star System	**Chi (χ) Cet** Multiple Star System
aka Baten Kaitos, 55 Cet, HR 539, HD 11353, SAO 148059, PPM 210734, HIP 8645, BD -11 00359, GC 2249, CCDM 1515-1019, NSV 638	aka 53 Cet, HR 531, HD 11171, SAO 148036, PPM 210688, HIP 8497, BD -11 00352, GC 2212
Suspected Variable	Magnitude: 4.68, F2
Mag: V 3.68 to V 3.74, K0III	Distance: 77 +/- 1.6 ly
Distance: 260 +/- 17.0 ly	AB: 4.66+6.8 mag, BUP, B=HD 11131
AB: 3.74+10.1 mag, BUP, B=HD 11366	In 1891: PA 250° Sep 183.80"
In 1907: PA 41° Sep 187"	

Object data from SKYTOOLS™ 2 — used with permission.

The Zeta – Chi (ζ-χ) Ceti Asterism / Cetus / ζ Cet / RA: 01h 51m 28s, Dec: -10° 20' 06"

Naked-Eye:

Eyepiece: 37x, 80.0'

Finder

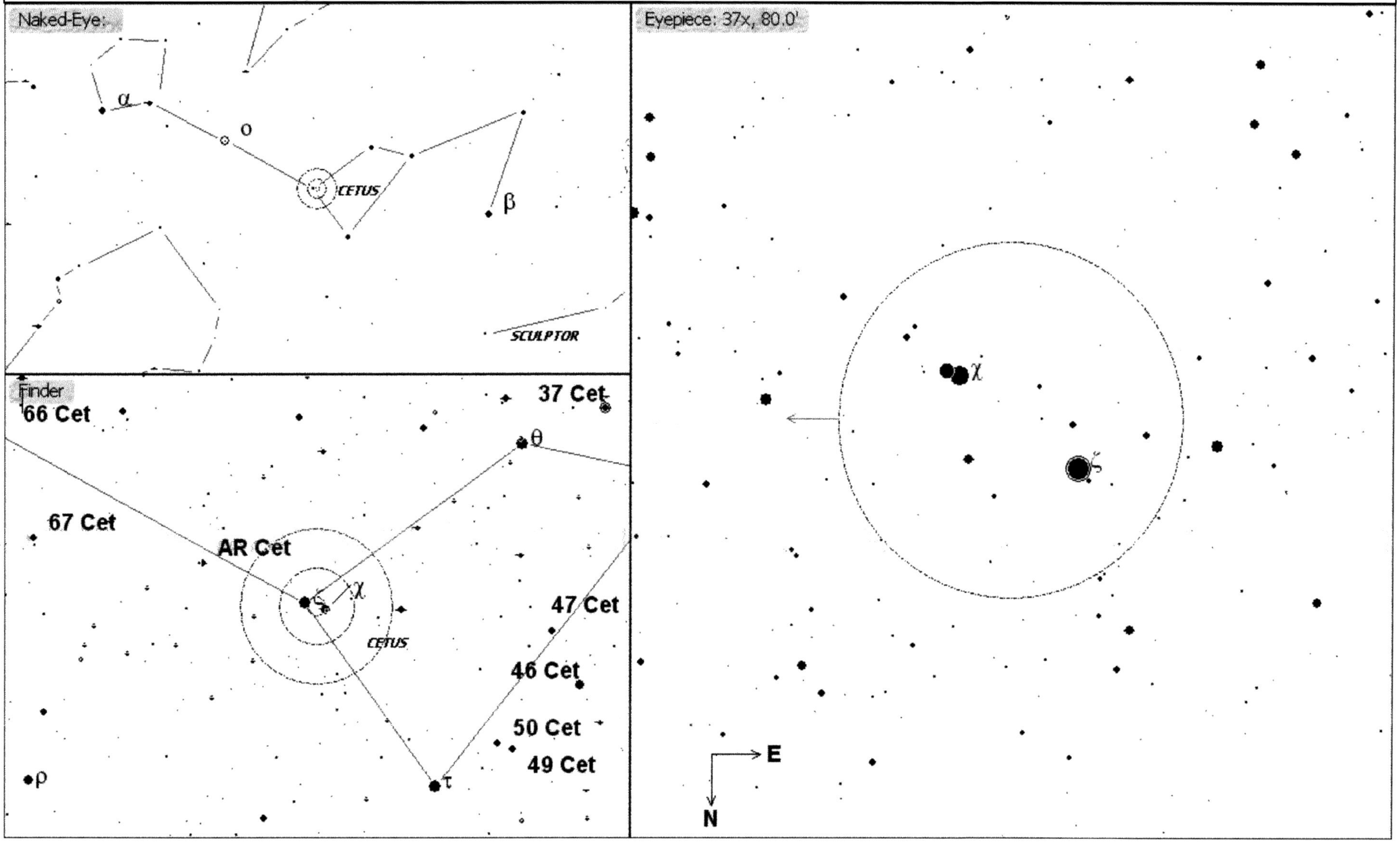

6" f/8 Reflector

Name	**The Upsilon (υ) Ceti Asterism**
Location	RA: 02h 00m 00s, Dec: -21° 04' 40" (2000) in Cetus
Found With	Direct Visual Observation
Colorful Stars	Upsilon and 57 Ceti
Description	The eyes see Upsilon and 57 Ceti together as a fuzzy spot in southern Cetus. The distance between the two stars is 15.5 arc minutes.

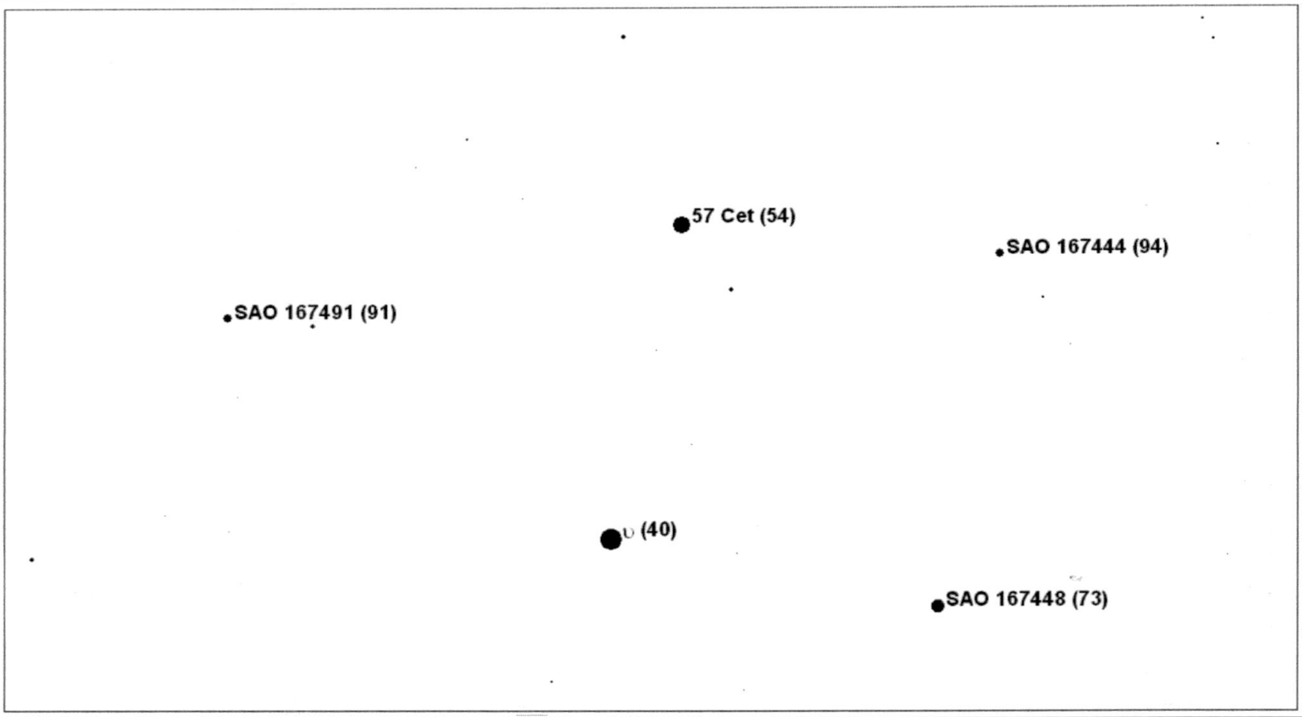

Principal Stars

57 Cet Single Star aka HR 583, HD 12255, SAO 167466 Magnitude: 5.43, M0III Distance: 670 +/- 110.0 ly	***Upsilon Cet** Single Star aka 59 Cet, HR 585, HD 12274, SAO 167471 Magnitude: 4.02, K5III Distance: 300 +/- 22.0 ly

Object data from SkyTools™ 2 — used with permission.

The Upsilon (υ) Ceti Asterism / Cetus / RA: 02h 00m 00s, Dec: -21° 04' 40"

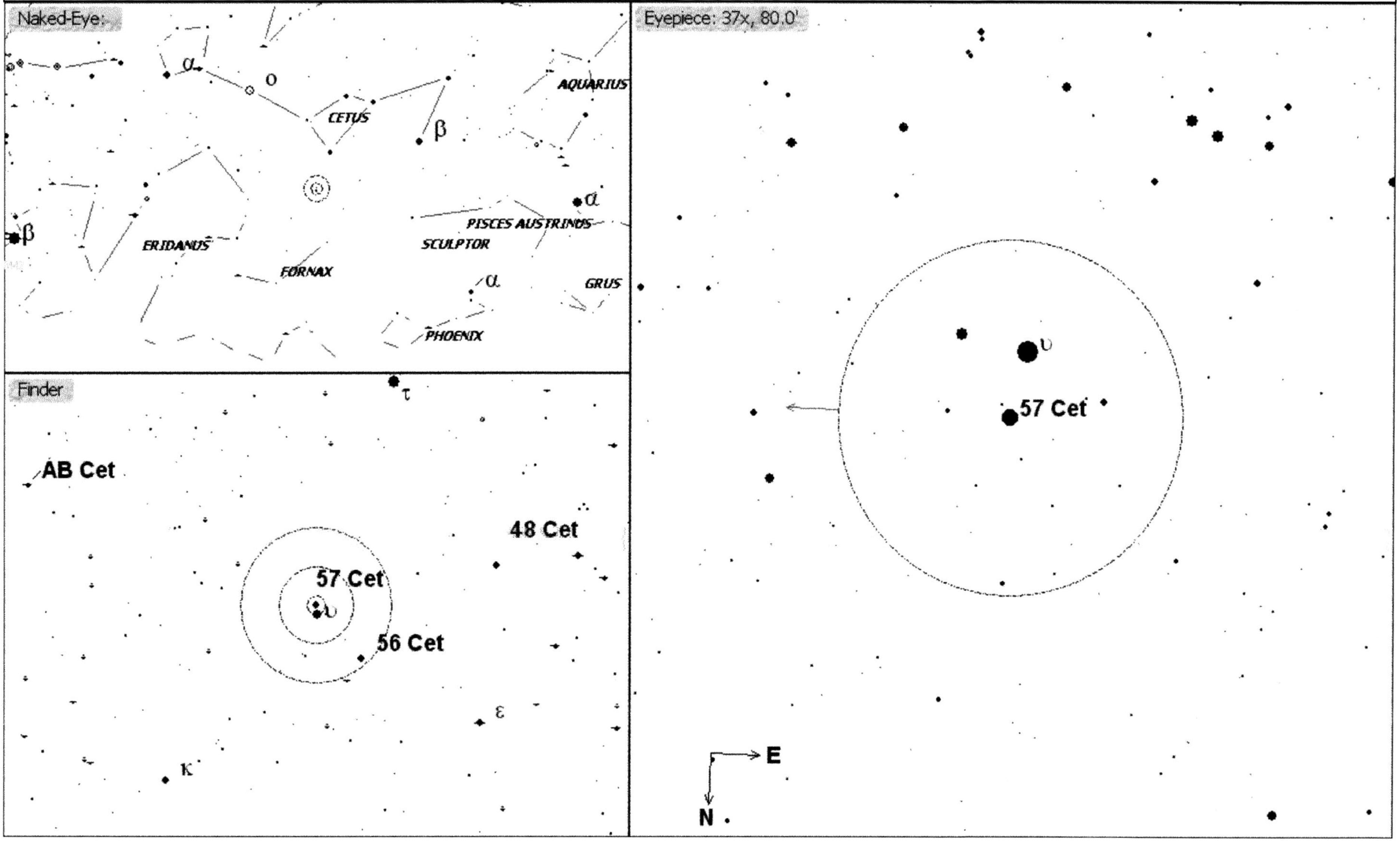

Naked-Eye

Eyepiece: 37x, 80.0'

Finder

6" f/8 Reflector

Name	**The Xi¹ (ξ¹) and 64 Ceti Asterism**
Location	RA: 02h 13m 00s, Dec: +08° 50' 48" (2000) in Cetus
Found With	Direct Observation
Description	A visual fuzzy object in northern Cetus. Binoculars reveal the two main stars distinctly and the string of fainter stars between them. The distance between Xi¹ and 64 is 29.5 arc minutes, almost one half degree.

Principal Stars

SAO 110383   Single Star
aka HD 13374, PPM 145353, HIP 10171
Magnitude: 8.15, G5
Distance: 1700 +/- 1300.0 ly

SAO 110388   Single Star
aka PPM 145358, BD +07 00346,
TYC 00630-1234 1
Magnitude: 9.67, G0

64 Cet   Single Star
aka HR 635, HD 13421, SAO 110390,
PPM 145360
Magnitude: 5.64, G0IV
Distance: 140 +/- 5.1 ly

TYC 00630-0692 1   Single Star
aka SI 54871
Magnitude: 10.27

SAO 110396 Single Star
aka HD 13484, PPM 145371
Magnitude: 9.07, A5

SAO 110404   Single Star
aka PPM 145382, BD +08 00343,
Magnitude: 8.81, K5

***Xi ¹ Cet**   Variable Star
aka 65 Cet, HR 649, HD 13611,
SAO 110408, PPM 145390, HIP 10324,
BD +08 00345, GC 2656, NSV 749,
Suspected Variable
Mag: V 4.35 to V 4.38, G8II
Distance: 360 +/- 51.0 ly

Object data from SKYTOOLS™ 2 — used with permission.

The Xi (ξ) and 64 Ceti Asterism / Cetus / ξ Cet / RA: 02h 13m 00s, Dec: +08° 50' 48"

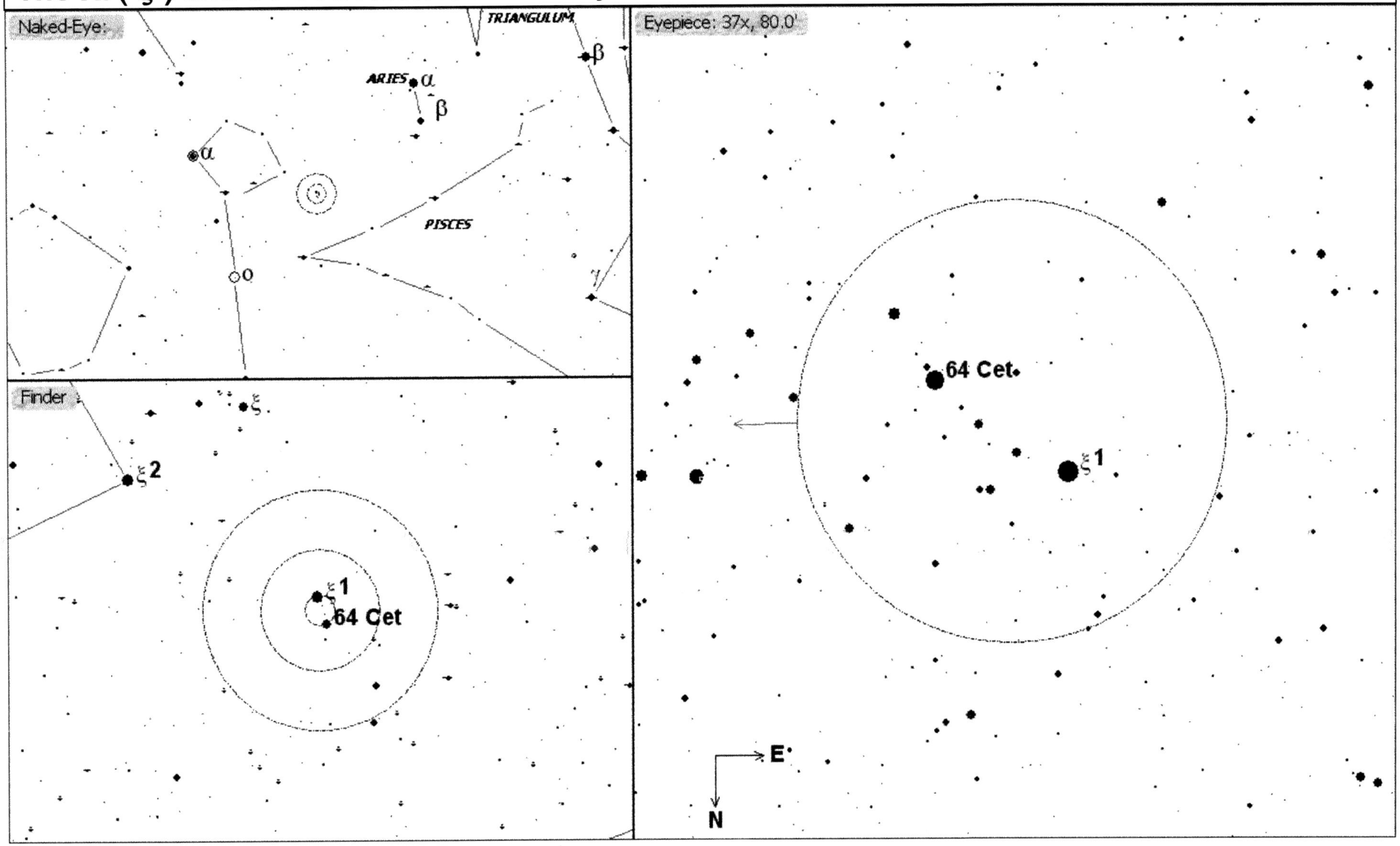

Eyepiece: 37x, 80.0'

Naked-Eye:

Finder:

6" f/8 Reflector

Name	**The 77 – 80 Ceti Asterism**
Location	RA: 02h 36m 00s, Dec: -07° 49' 54" (2000) in Cetus
Found With	Direct Vision
Colorful Stars	77 & 80 Cet
Description	Another wide pair of stars that appears as a fuzzy object in the sky. The distance between the 77 and 80 is 19 arc-minutes. The B component of each star is a challenge for small scopes.

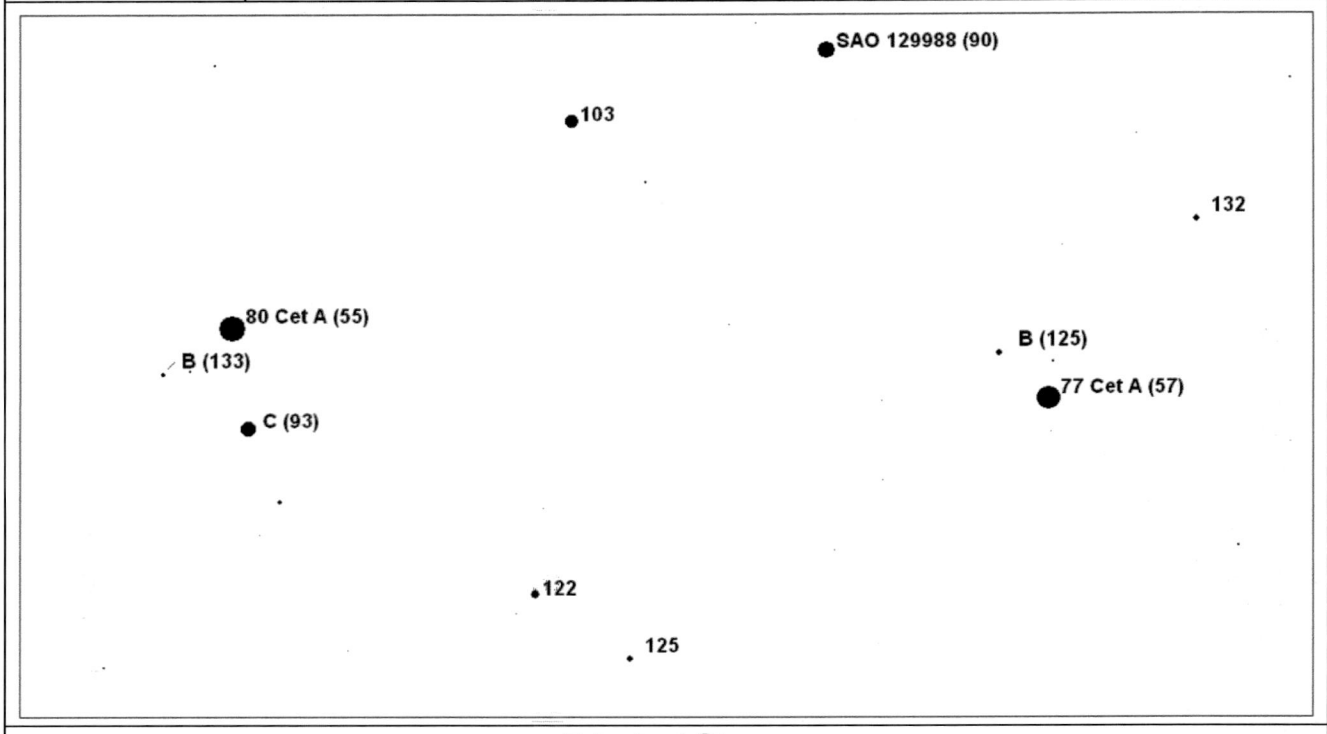

(star chart showing SAO 129988 (90), 103, 132, 80 Cet A (55), B (133), C (93), B (125), 77 Cet A (57), 122, 125)

Principal Stars	
*80 Cet Multiple Star System aka HR 759, HD 16212, SAO 130004, PPM 184807, HIP 12107, BD -08 00489, BUP Magnitude: 5.53, K5 Distance: 520 +/- 71.0 ly AB: 5.53+13.3 mag In 1911: PA 123° Sep 117.30" AC: 5.53+9.3 mag, BUP C=SAO 130003 In 1900: PA 189° Sep 146.70"	**77 Cet** Multiple Star System aka HR 752, HD 16074, SAO 129984, PPM 184780, HIP 12002, BD -08 00484, BUP Magnitude: 5.74, K0 Distance: 550 +/- 81.0 ly AB: 5.74+12.5 mag In 1911: PA 47° Sep 95.10"

Object data from SKYTOOLS™ 2 — used with permission.

The 77 – 80 Ceti Asterism / Cetus / 80 Ceti / RA: 02h 36m 00s, Dec: -07° 49' 54"

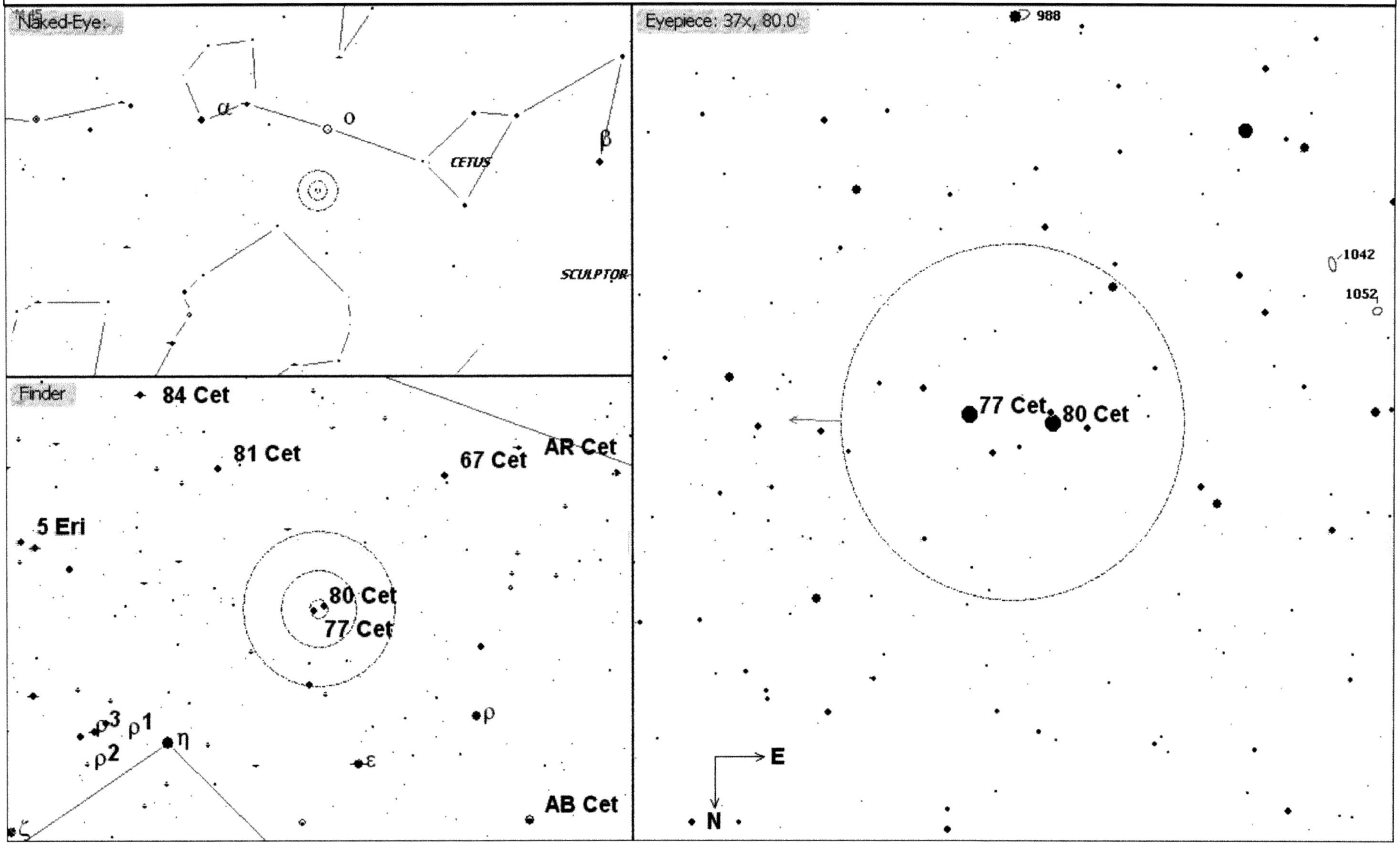

Naked-Eye:

Finder:

Eyepiece: 37x, 80.0'

6" f/8 Reflector

Name	**The Delta ^{2,3} (δ ^{2,3}) Canis Minoris Asterism**
Location	RA: 07h 33m 12s, Dec: +03° 17' 25" (2000) in Canis Minor
Found With	Eyes : Seen with Averted Vision Only
Description	A very faint fuzzy spot seen only with averted vision. Binoculars or a small telescope reveal a group of four stars about 2 degrees southwest of Procyon. The distance between Delta 2 and Delta 3 is about 17'.

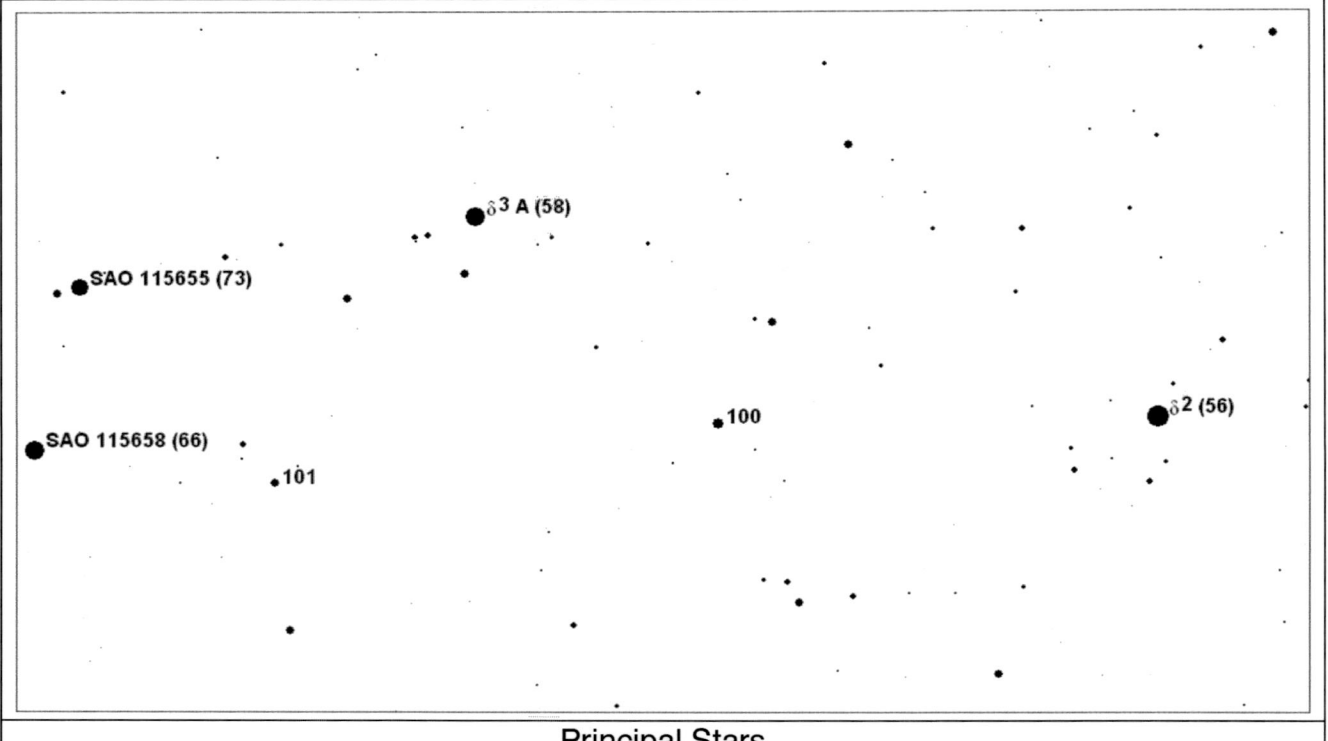

Principal Stars

*Delta 2 Cmi Single Star
aka 8 CMi, HR 2887, HD 60111,
SAO 115610, PPM 152860, HIP 36723
Magnitude: 5.59, F2
Distance: 140 +/- 4.6 ly

Delta 3 Cmi Multiple Star System
aka 9 CMi, HR 2901, HD 60357,
SAO 115644, HIP 36812, GC 10128
Magnitude: 5.83, A0
Distance: 680 +/- 110.0 ly
AB: 5.83+11 mag
In 1945: PA 110° Sep 90.00"
BC: 11+13 mag, J 2835
In 1945: PA 195° Sep 5.00"

SAO 115655 Single Star
aka HD 60504, PPM 152922, HIP 36879
Magnitude: 7.25, G5
Distance: 500 +/- 76.0 ly

SAO 115658 Single Star
aka HD 60526, PPM 152924, HIP 36882
Magnitude: 6.56, G5
Distance: 510 +/- 69.0 ly

Object data from SKYTOOLS™ 2 — used with permission.

The Delta ^{2,3} ($\delta^{2,3}$) Canis Minoris Asterism / Canis Minor / δ^2 Cmi / RA: 07h 33m 12s, Dec: +03° 17' 25"

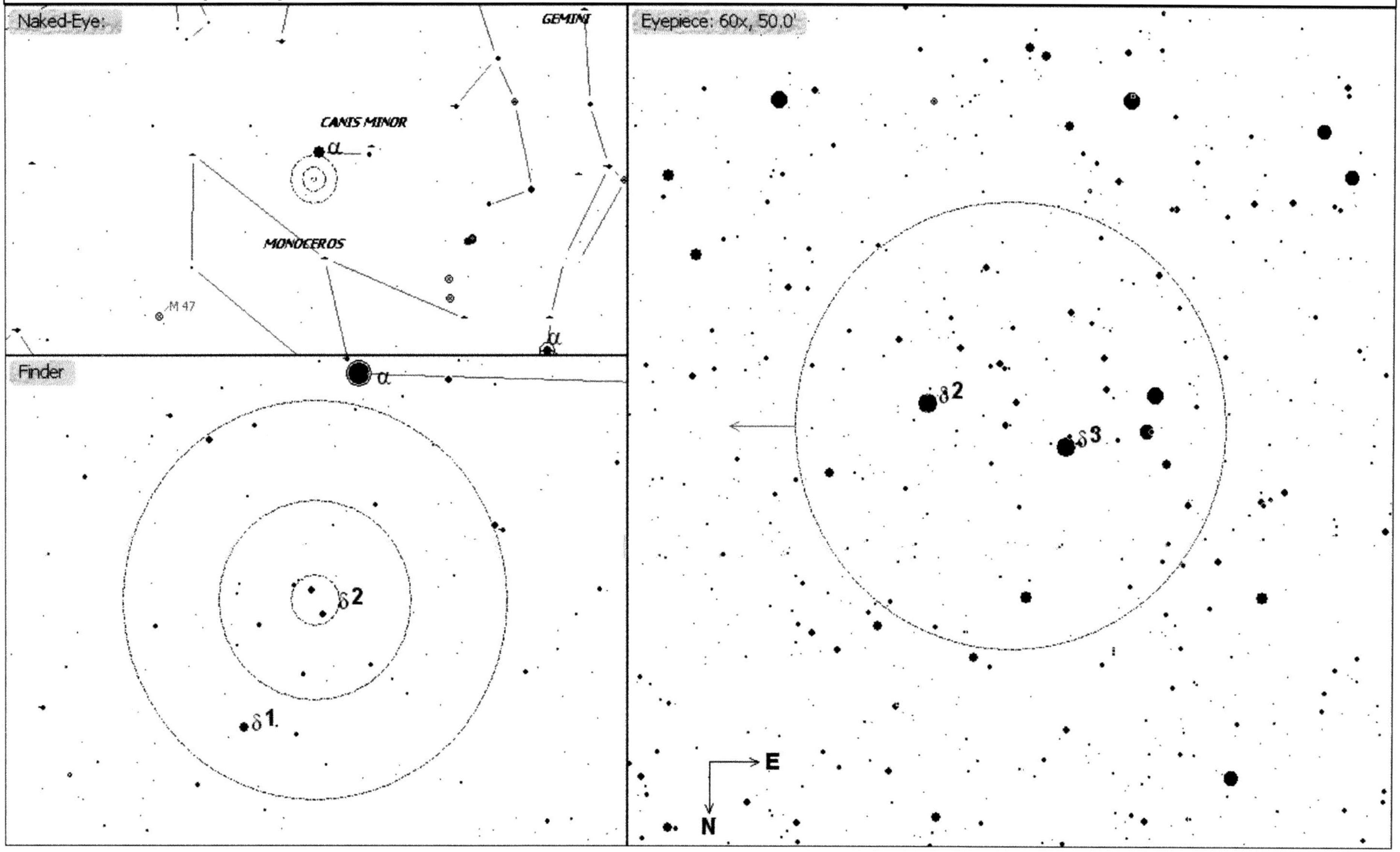

Eyepiece: 60x, 50.0

Naked-Eye

Finder

E, N

6" f/8 Reflector

Name	**The Cancer Cascade Asterism**
Location	RA: 08h 00m 57s, Dec: +25° 23' 34" (2000) in Cancer
Found With	Binoculars
Description	This faint stream of stars reminds me of the brighter Kemble's Cascade in Camelopardalis. The cascade starts with Omega[1] Cancri and ends two degrees later at SAO 79961. Its best seen at low power.

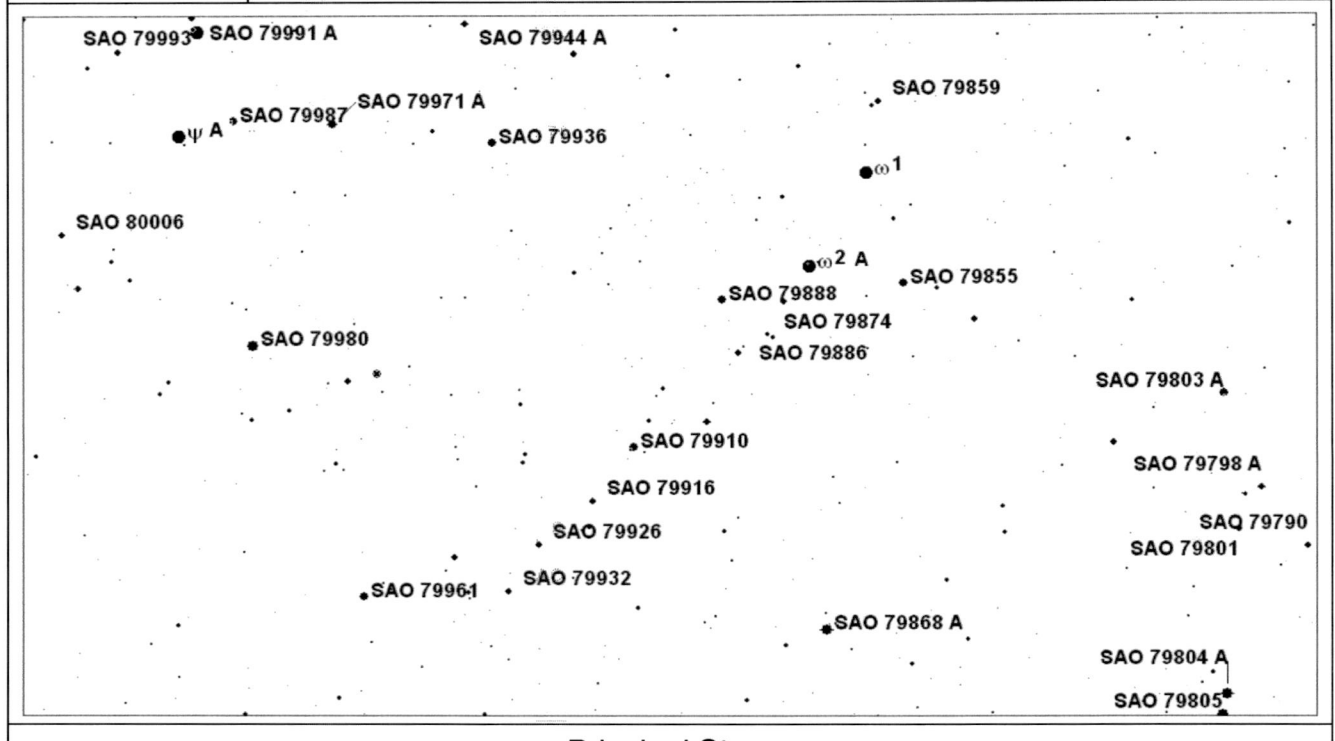

Principal Stars

*Omega ^{1} (ω^1) Cnc   Single Star
aka 2 Cnc, HR 3124, HD 65714, SAO 79861
Magnitude: 5.87, K0
Distance: 1100 +/- 340.0 ly

Omega^2 (ω^2) Cnc   Multiple Star System
aka 4 Cnc, HR 3132, HD 65856, SAO 79869, PPM 98236, HIP 39263, H 75
Magnitude: 6.32, F5, Distance: 600 +/- 100.0 ly
AB: 6.3+11 mag, In 1904: PA 23° Sep 45.5"
AC: 6.3+10 mag, In 1904: PA 295° Sep 109.3"

SAO 79874   Single Star
aka PPM 98244, HIP 39297, BD +25 01819
Mag: H 8.60 to H 8.77, K7
Distance: 4500 +/- 7300.0 ly

SAO 79888   Single Star
aka HD 66086, PPM 98263, BD +25 01826
Magnitude: 8.19, K5

SAO 79886   Single Star
aka HD 66047, PPM 98257, BD +25 01825
Magnitude: 8.58, K0

SAO 79910   Single Star
aka HD 66371, PPM 98300, BD +24 01840
Magnitude: 8.01, M0

SAO 79892   Multiple Star System
aka HD 66140, PPM 98268, HIP 39387
Magnitude: 8.94, F0, Distance: 980 +/- 500.0 ly
AB: 8.94+12.36 mag; In 1991: PA 10° Sep 0.55"

SAO 79916   Single Star
aka HD 66485, PPM 98307, HIP 39528
Magnitude: 8.43, G5
Distance: 140 +/- 7.4 ly

SAO 79926   Single Star
aka HD 66663, PPM 98323, BD +24 01847
Magnitude: 8.67, G5

SAO 79932   Single Star
aka HD 66753, PPM 98333, BD +24 01852
Magnitude: 8.76, F5

SAO 79949   Single Star
aka HD 66921, PPM 98352, HIP 39686
Magnitude: 8.97, F5
Distance: 490 +/- 86.0 ly

SAO 79961   Single Star
aka HD 67226, PPM 98372, HIP 39790
Magnitude: 8.06, K0
Distance: 1100 +/- 380.0 ly

Object data from SKYTOOLS™ 2 — used with permission.

The Cancer Cascade Asterism / Cancer / Omega¹ (ω¹) Cancri / RA: 08h 00m 56s, Dec: +25° 23' 34"

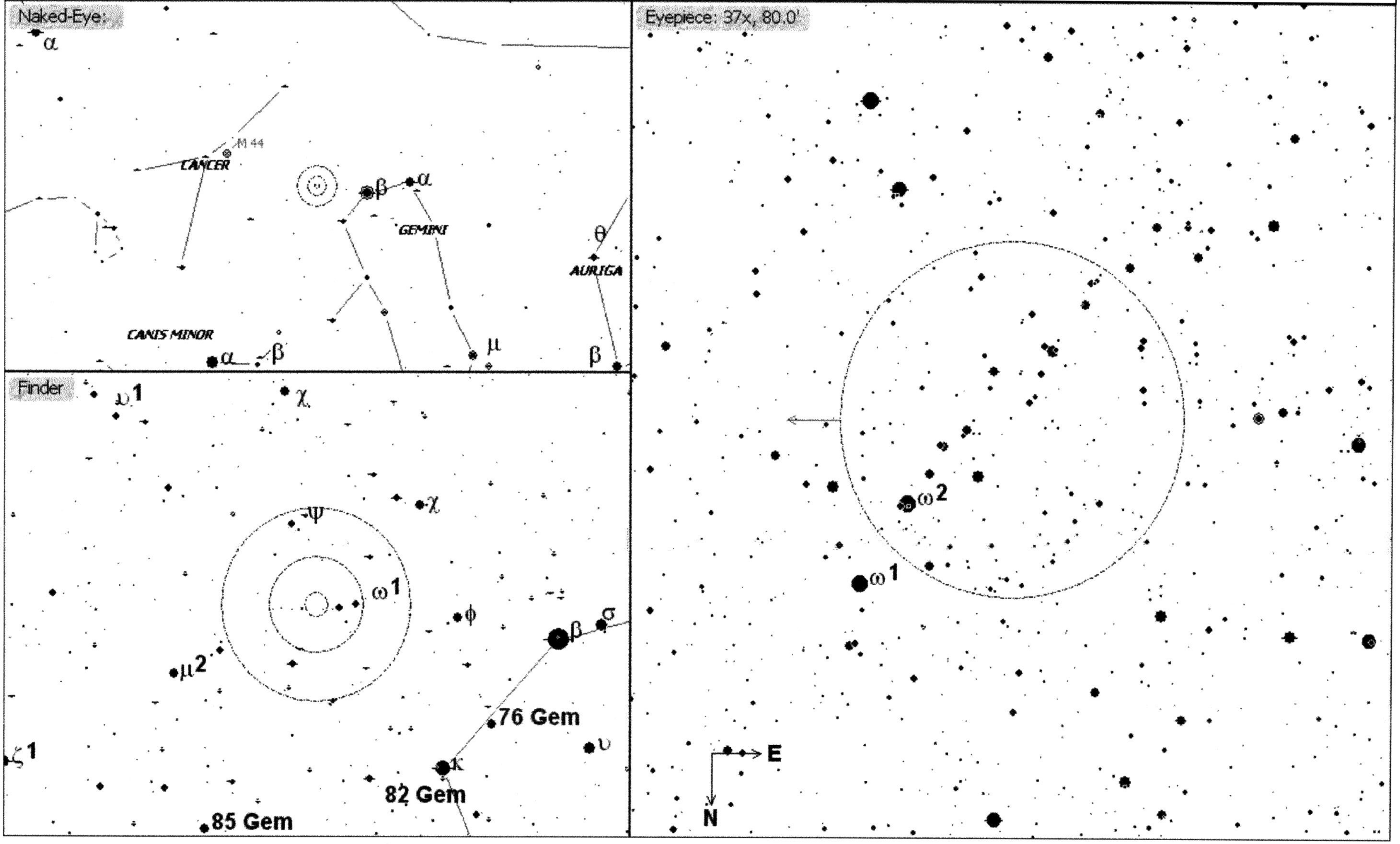

6" f/8 Reflector

Name	**The Sigma³ (σ³) Cancri Asterism**
Location	RA: 08h 59m 33s, Dec: +32° 25' 07" (2000) in Cancer
Found With	Direct Observation
Description	This asterism can be seen with the naked eye as a fuzzy spot in northern Cancer, about 13 degrees north of the Beehive. Optical aid shows a group of stars in a one-degree field. The distance from Sigma² to Sigma³ is 45 arc-minutes; from Sigma³ to 66 Cnc is 25 arc min.

Principal Stars

Sigma 2 Cnc Single Star
aka 59 Cnc, HR 3555, HD 76398, SAO 61146
Magnitude: 5.44, A7IV
Distance: 190 +/- 11.0 ly

SAO 61151 Multiple Star System
aka PPM 74122, Struve 1294
Magnitude: 8.87, K2
AB: 8.79+10.9 mag, In 1830: PA 340° Sep 15"

SAO 61152 Single Star
aka HD 76461, PPM 74123, HIP 43976
Magnitude: 6.99, A5
Distance: 270 +/- 21.0 ly

SAO 61163 Multiple Star System
aka HD 76676, PPM 74137, BD +32 01816
Magnitude: 7.18, K5
AB: 7.1+8.7 mag, In 1955: PA 246° Sep 0.30"

SAO 61169 Single Star
aka HD 76765, PPM 74143, HIP 44118
Magnitude: 7.95, G5
Distance: 200 +/- 12.0 ly

SAO 61188 Single Star
aka HD 76945, PPM 74163, BD +32 01825
Magnitude: 7.59, F2

Object data from SKYTOOLS™ 2 — used with permission.

*Sigma 3 Cnc Multiple Star System
aka 64 Cnc, HR 3575, HD 76813,
SAO 61177, SHJ 100
Magnitude: 5.23, G8III, Distance: 320 +/- 24.0 ly
AB: 5.23+9.4 mag, , B=SAO 61176
In 1823: PA 295° Sep 89.60"

SAO 61192 Multiple Star System
aka HD 76975, PPM 74167, HIP 44230,
ADS 7124, HU 718
Magnitude: 8.51, G5
Distance: 350 +/- 56.0 ly
AB: 8.51+9.49 mag, PA 53° Sep 0.34" (2005.3)
Preliminary Orbit: P=294.0 yr, a=0.56"

SAO 61198 Single Star
aka HD 77007, PPM 74170, BD +32 01827
Magnitude: 8.45, K0

66 Cnc Multiple Star System
aka HR 3587, HD 77104, SAO 61202,
Struve 1298, ADS 7137
Magnitude: 5.89, A2
Distance: 510 +/- 72.0 ly
AB: 5.89+8.79 mag, , B=HIP 44307
In 1991: PA 135° Sep 4.52"
AC: 5.89+10.8 mag,
In 1879: PA 319° Sep 187.40"

The Sigma³ (σ³) Cancri Asterism / Cancer / RA: 08h 59m 33s, Dec: +32° 25' 07"

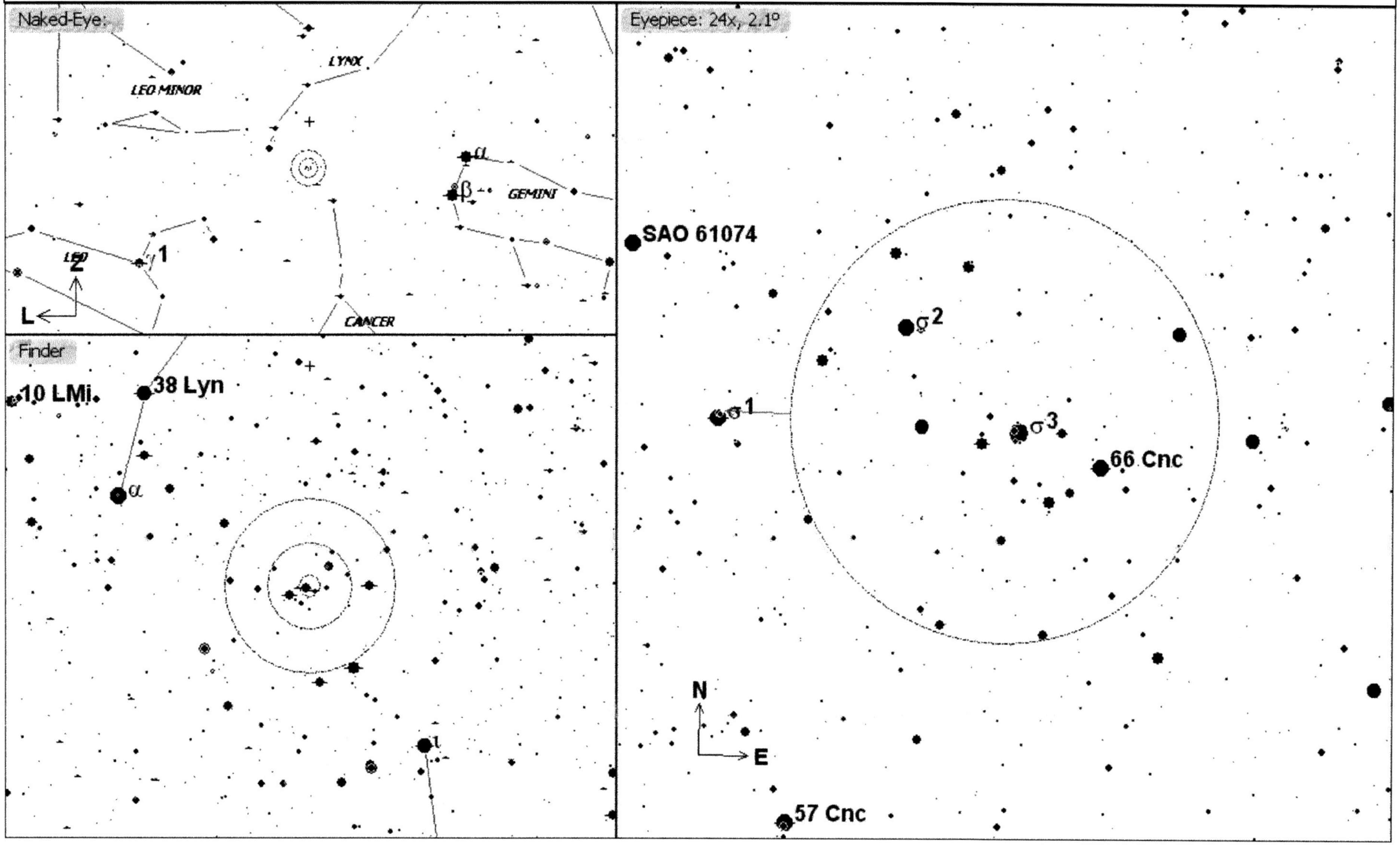

Eyepiece: 24x, 2.1°

Naked-Eye:

Finder:

σ2
σ3
σ1
66 Cnc
57 Cnc
SAO 61074

N E

38 Lyn
10 LMi
α

LEO MINOR
LYNX
CANCER
GEMINI

80mm f6 Refractor

75

Name	**The M53 Asterism**
Location	RA: 13h 13m 12s, Dec: +18° 43' 37" (2000) in Coma Berenices
Colorful Stars	SAO 100476, 100489, 100496, 100499
Found With	12.5" Telescope
Description	This telescopic asterism is one half degree north of Messier 53 and stretches east for another degree. Five of its members are brighter than magnitude 7.0, an impressive number to be within one square degree of sky. On the finder chart is plotted NGC 5053, one degree to the SE of M53 at PA 120°. NGC 5053 is a challenge object even in larger scopes. Its appears much fainter than its listed magnitude would suggest.

Principal Stars

SAO 100459 Single Star
aka HD 114745, PPM 129657, HIP 64429,
Magnitude: 8.19, F8
Distance: 440 +/- 89.0 ly

SAO 100461, Single Star
aka HD 114793, HR 4987, PPM 129660
Magnitude: 6.51, G8III

*SAO 100467 Single Star
aka HR 4992, HD 114889, PPM 129669, HIP 64496
Magnitude: 6.10, G8III
Distance: 300 +/- 32.0 ly

SAO 100476   Single Star
aka HD 115166, PPM 129686, HIP 64654
Magnitude: 6.55, K0
Distance: 610 +/- 97.0 ly

SAO 100484   Single Star
aka HR 5007HD 115319, PPM 129696, HIP 64751
Magnitude: 6.46, G8III
Distance: 310 +/- 23.0 ly

SAO 100487 Single Star
aka HD 115351, PPM 129700, BD +19 02656
Magnitude: 8.45, G0
Distance: 420 +/- 43.0 ly

SAO 100489   Single Star
aka HD 115381, PPM 129703, HIP 64781
Magnitude: 6.72, K0

SAO 100496 Single Star
aka HD 115464, PPM 129712, BD +20 02816
Magnitude: 8.50, K0

SAO 100499 Single Star
aka HD 115537, PPM 129717, HIP 64849
Magnitude: 7.93, K2
Distance: 900 +/- 260.0 ly

SAO 100502 Single Star
aka HD 115558, PPM 129718, HIP 64855
Magnitude: 8.40, G5
Distance: 1600 +/- 800.0 ly

M 53   Globular Cluster
aka NGC 5024
Magnitude: 7.70
Size: 13.0', Distance: 65000 ly

NGC 5053   Globular Cluster (On finder chart)
Magnitude: 9.00
Size: 10.0',
Distance: 55000 ly

Object data from SKYTOOLS™ 2 — used with permission.

The M53 Asterism / Coma Bereneces / SAO 100467 / RA: 13h 13m 12s, Dec: +18° 43' 37"

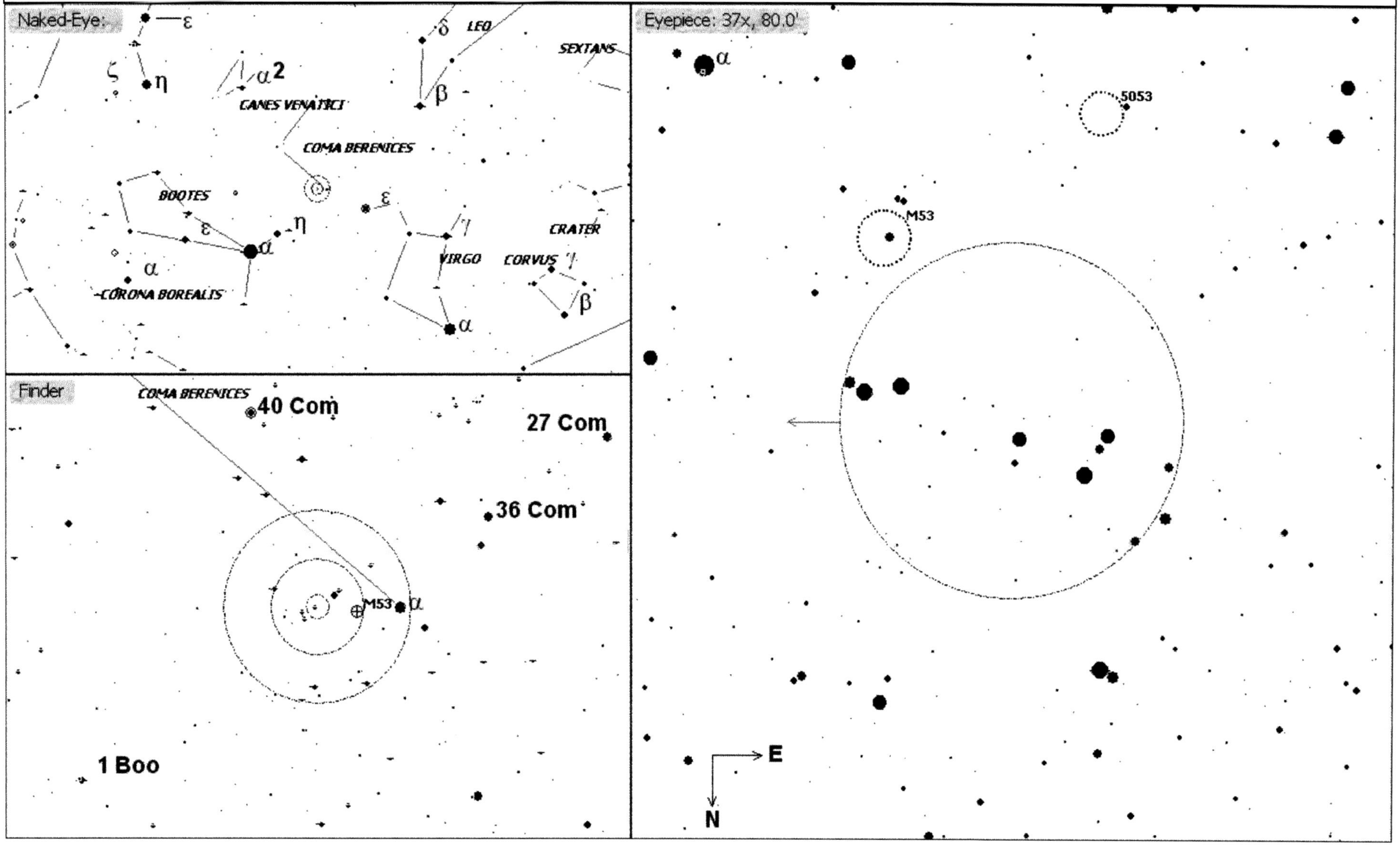

Eyepiece: 37x, 80.0'

Naked-Eye

Finder

6" f/8 Reflector

Name	**The AW Canum Venaticorum Asterism**
Location	RA: 13h 51m 48s, Dec: +34° 26' 39" (2000) in Canes Venatici
Found With	8 x 50 Finderscope
Colorful Stars	AW CVn, SAO 63781
Description	This asterism is quite noticeable in binoculars as a fuzzy object. A telescope reveals the ruddy color of AW and SAO 63781. The asterism is further enhanced by the faint double star SAO 63785. The distance between AW and SAO 63779 is 21 arc-min. This is a real treat for larger apertures.

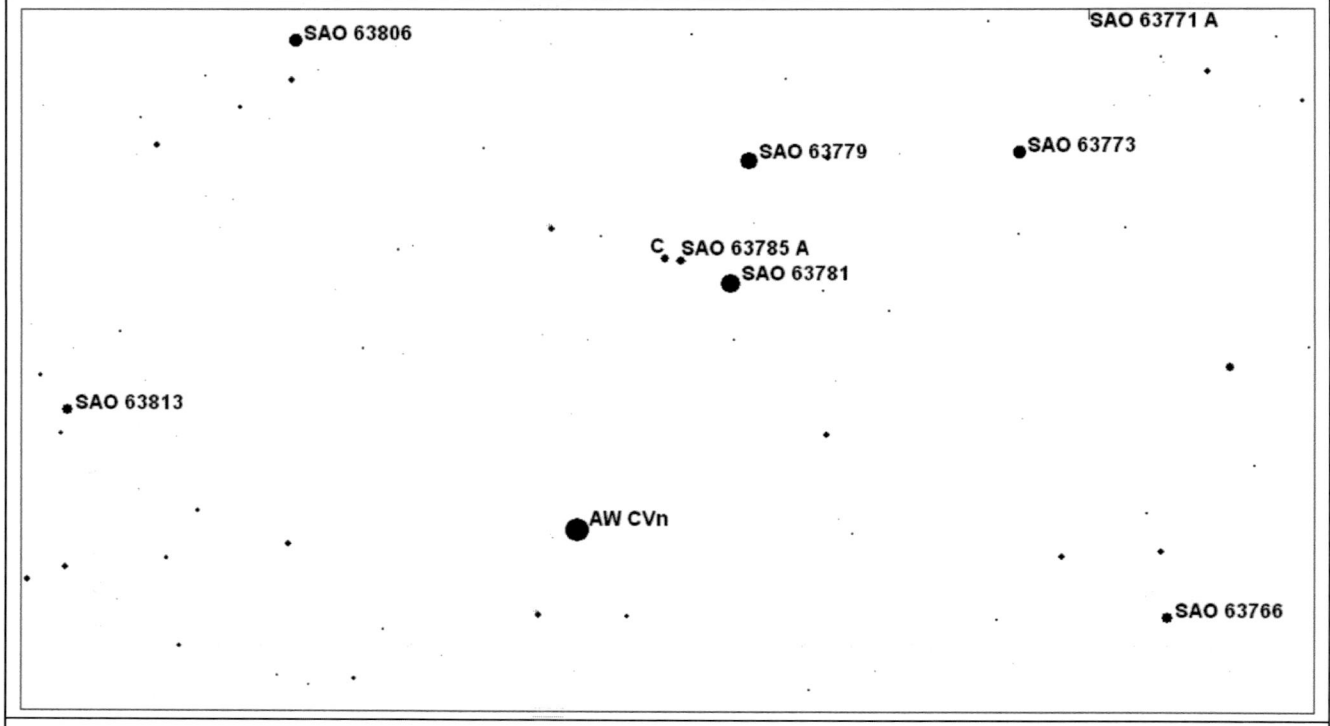

Principal Stars

SAO 63779 Single Star
aka HD 120818, HR 5214, PPM 77470, HIP 67596
Magnitude: 6.65, A5IV
Distance: 290 +/- 18.0 ly

SAO 63781 Variable Star
aka HD 120819, HR 5215, PPM 77472, HIP 67605, BD +35 02493, GC 18726
Suspected Variable
Mag: V 5.87 to V 5.95, M2II
Distance: 680 +/- 98.0 ly

SAO 63785 Multiple Star System
aka PPM 77476, BD +35 02494, ADS 9037, BU 613
Magnitude: 10.29
AB: 10.15+10.9 mag
In 1878: PA 149° Sep 0.80"
AC: 10.15+10.3 mag, C=SAO 63786
In 1892: PA 83° Sep 48.70"

***AW CVn** Variable Star
aka GCVS 3835, HR 5219, HD 120933, SAO 63793, PPM 77486, HIP 67665
Variability Type: SR
Mag: V 4.72 to V 4.81, M0
Distance: 600 +/- 75.0 ly

Object data from SKYTOOLS™ 2 — used with permission.

The AW CVn Asterism / Canes Venatici / RA: 13h 51m 48s, Dec: +34°26' 39"

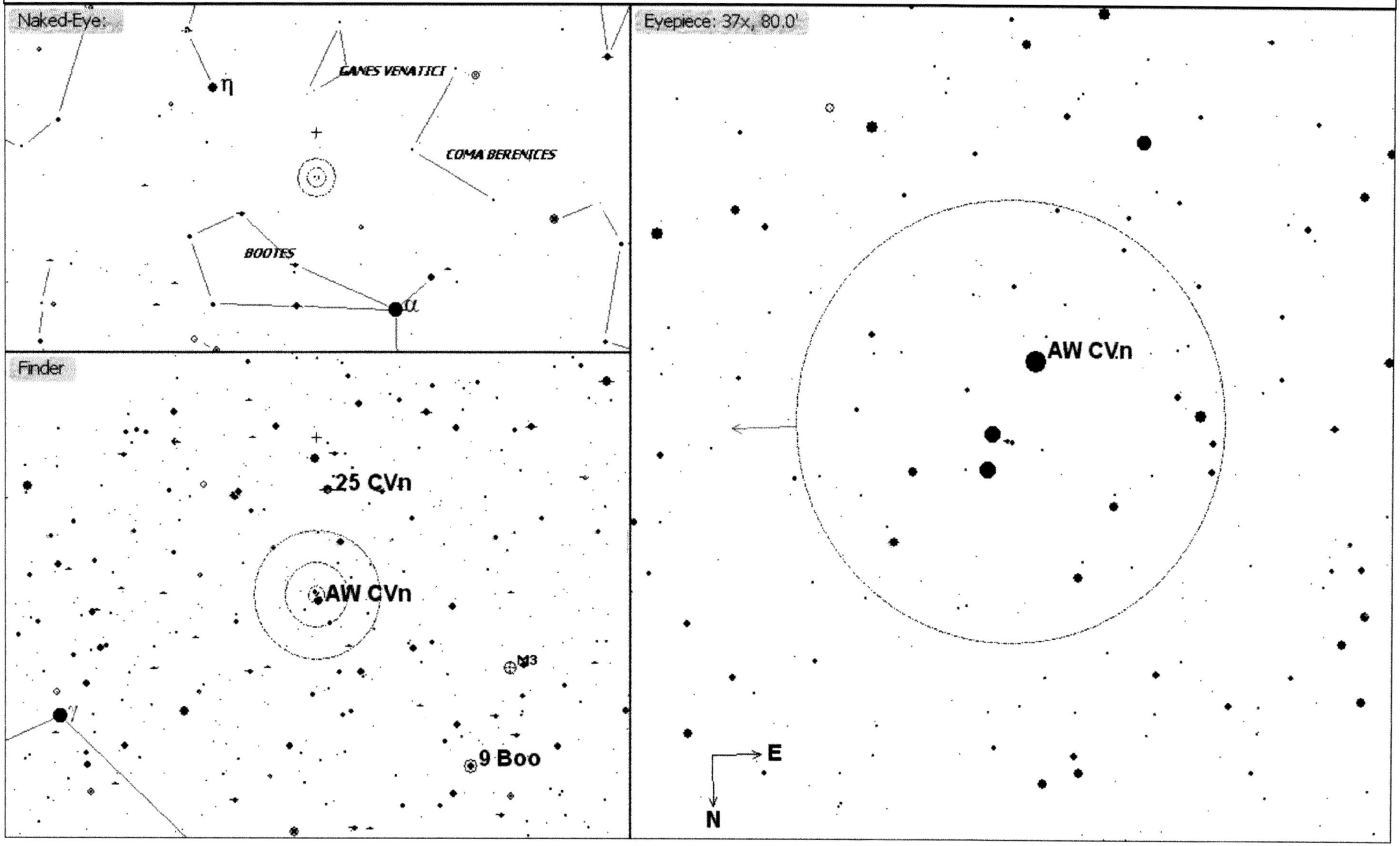

Eyepiece: 37x, 80.0'

E / N

Naked-Eye:

Finder

6" f/8 Reflector

Name	**The Struve 1659 Asterism**
Location	RA: 12h 35m 44s, Dec: -12° 01' 30" (2000) in Corvus
Found With	8 x 50 finderscope
Description	A neat multiple star one degree southwest of M104, seen as a fuzzy spot in the finderscope. The outer three stars average 3 arc-minutes distance from the center of the asterism.

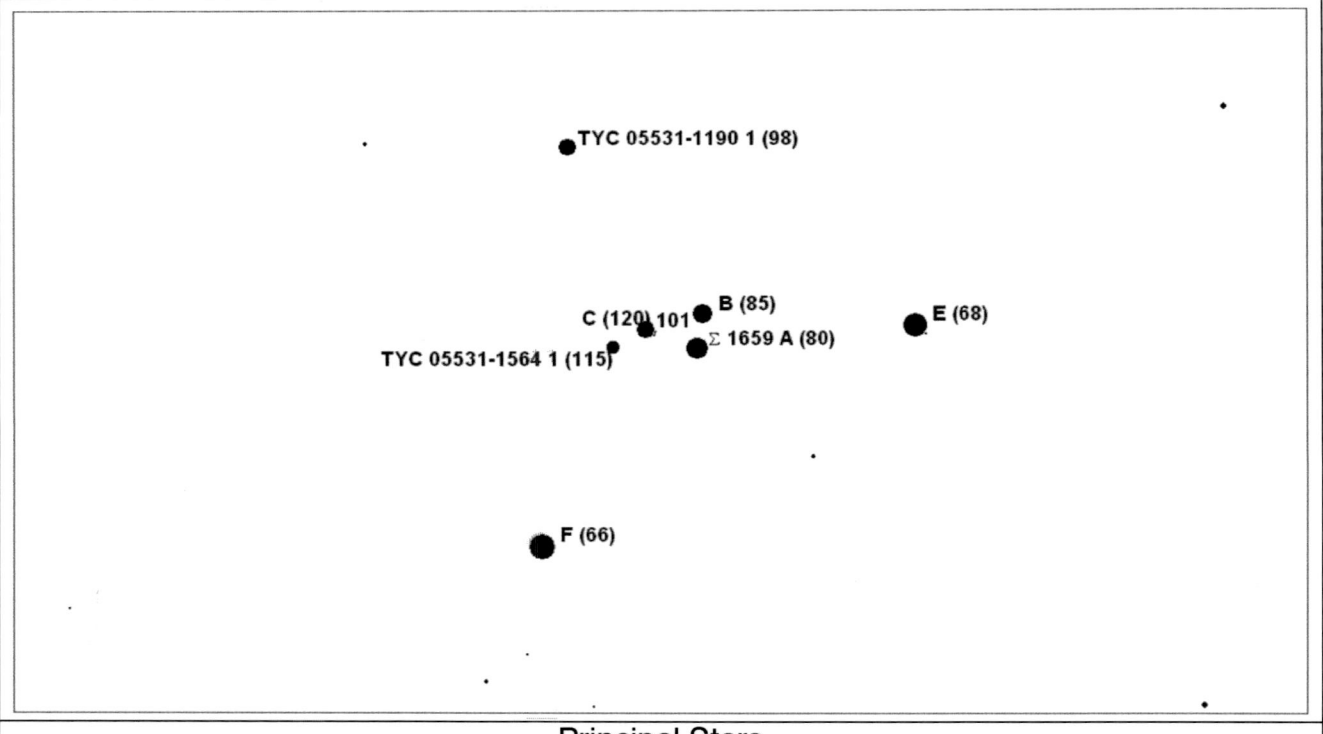

Principal Stars

Struve 1659 Multiple Star System
aka SAO 157384, HD 109556, PPM 226074
Component 'A' : Magnitude: 8.0, G0
PPM 717043 Single Star
(Very Close to Σ1659 'C')
aka SI 1070267
Magnitude: 10.10

TYC 05531-1564-1 Single Star
aka SI 581832
Magnitude: 11.5
TYC 05531-1190-1 Single Star
aka PPM 717044, SI 581814
Magnitude: 9.8

Σ 1659 Component Data

Components	Magnitudes	PA and Separation	Other Designation
AB	8 + 8.5	In 1991: PA 351° Sep 27.48"	B=SAO 157383
AC	8 + 12	In 1922: PA 70° Sep 37.80"	
AE	8 + 6.8	In 1908: PA 276° Sep 169.20"	E=HD 109545
AF	8 + 6.6	In 1870: PA 143° Sep 199.40"	F=HD 109584

Object data from SKYTOOLS™ 2 — used with permission.

The Struve 1659 Asterism / Corvus / RA: 12h 35m 44s, Dec: -12° 01' 30"

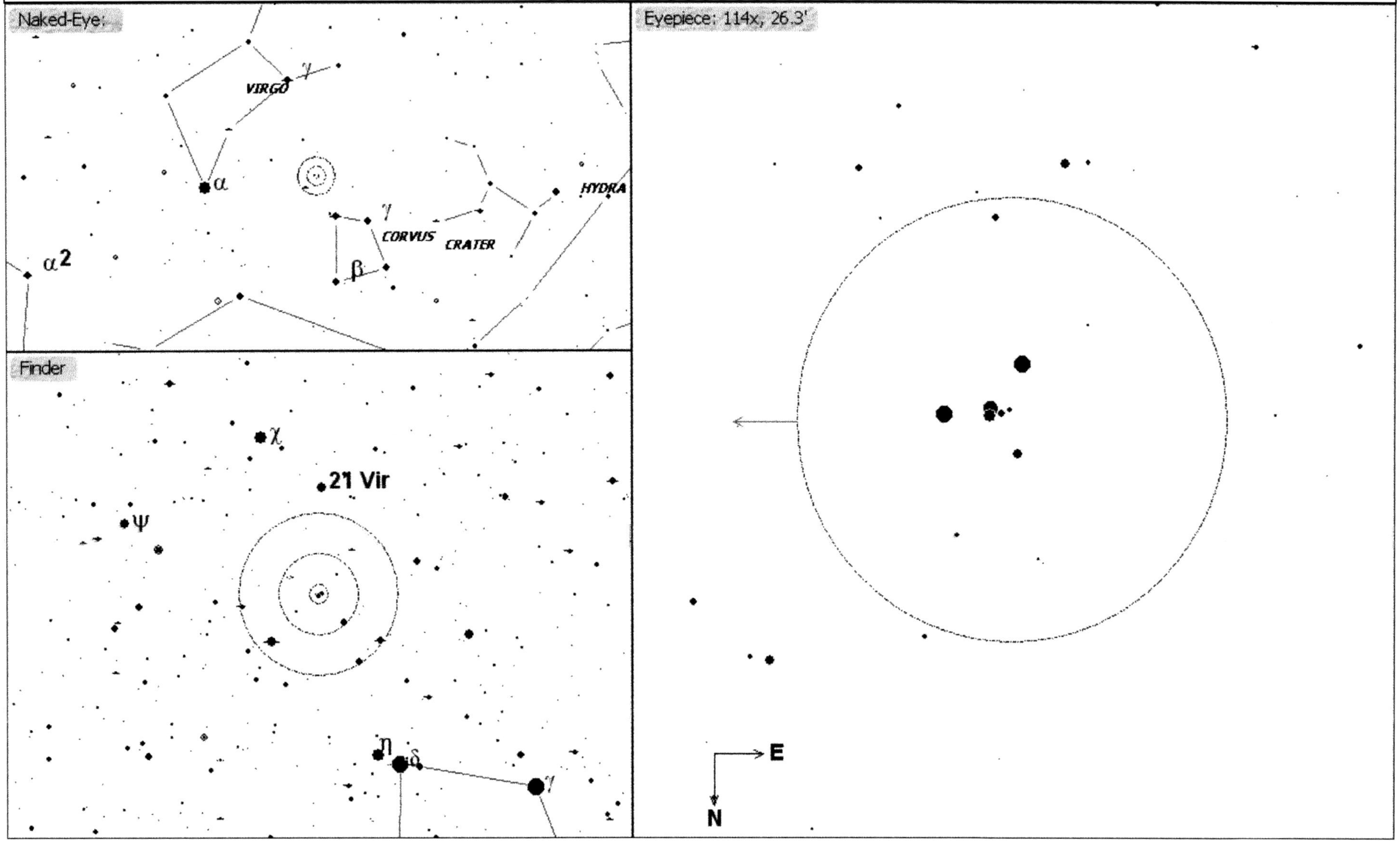

Naked-Eye

Finder

Eyepiece: 114x, 26.3'

6" f/8 Reflector

Name	**The Tau (τ) Coronae Borealis Asterism**
Location	RA: 16h 08m 58s, Dec: +36° 29' 27" (2000) in Corona Borealis
Found With	Direct Visual Observation
Description	This chain of four stars can be seen from dark skies. It appears as a faint, but perfectly straight east-west line of stars, 2.5 degrees in length. Extend this line 6 degrees east and it will pass less than one degree south of M13 in Hercules.

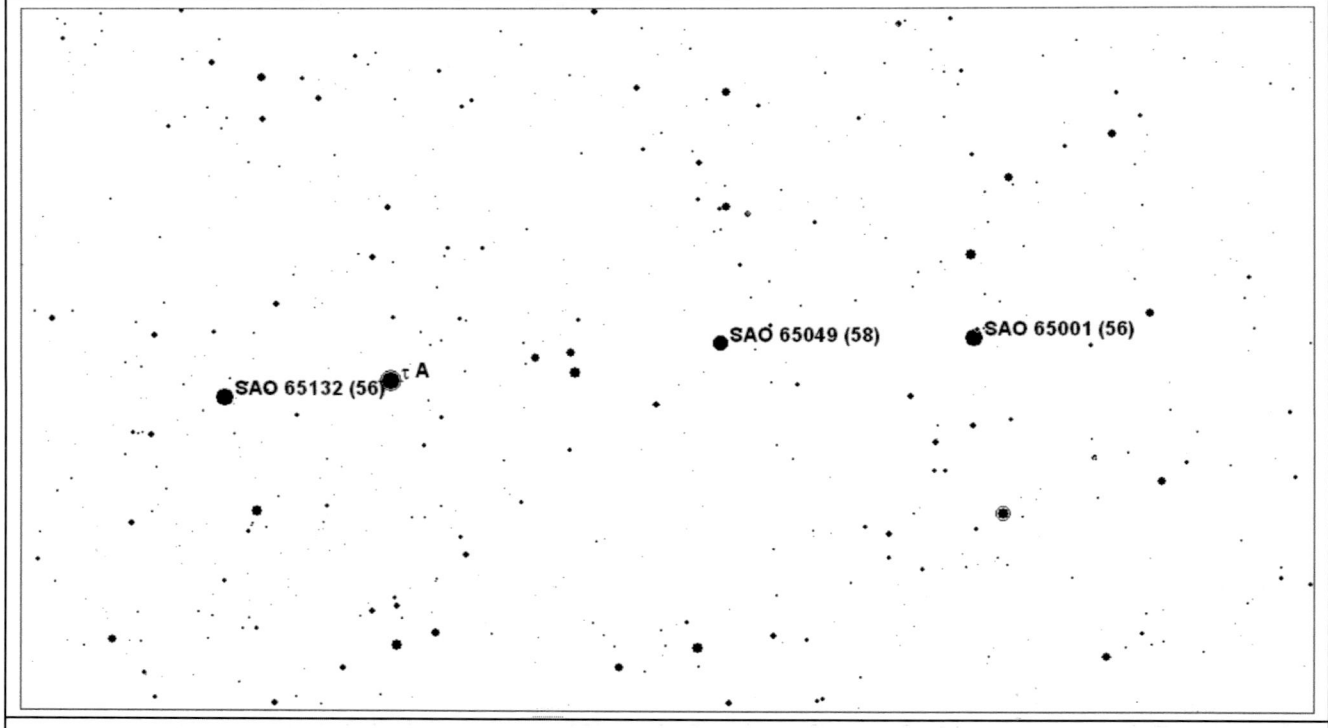

Principal Stars

SAO 65001 Single Star

aka HD 143435, HR 5957, PPM 78947

Magnitude: 5.61, K5

Distance: 900 +/- 140.0 ly

SAO 65049 Single Star

aka HD 144208, HR 5983, PPM 79002

Magnitude: 5.79, F5I

Distance: 690 +/- 86.0 ly

*Tau CrB Multiple Star System

aka 16 CrB, HR 6018, HD 145328, SAO 65108, BU 1087, ADS 9939

Suspected Variable

Mag: V 4.72 to V 4.82, K1III

Distance: 110 +/- 2.1 ly

AB: 4.73+13.2 mag

In 1958: PA 186° Sep 2.20"

SAO 65132   Multiple Star

aka HD 145849, HR 6046, PPM 79111, WRH 21

Magnitude: 5.62, K5

Distance: 610 +/- 69.0 ly

Sep: 0.1" (Not in My Backyard!!!)

Object data from SKYTOOLS™ 2 — used with permission.

The Tau (τ) Coronae Borealis Asterism / Corona Borealis / RA: 16h 08m 58s, Dec: +36° 29' 27"

Naked-Eye:

BOOTES
CORONA BOREALIS
HERCULES

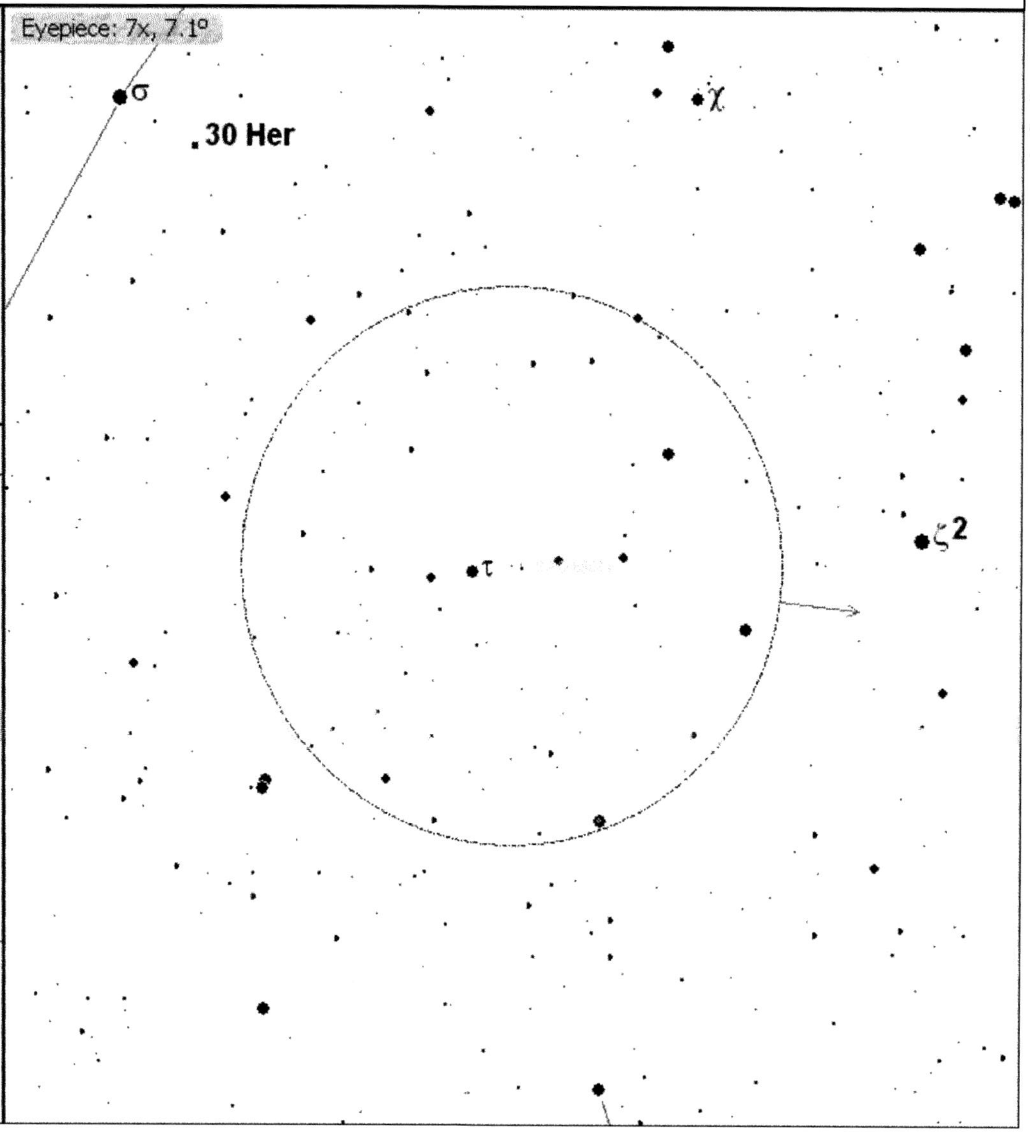

Eyepiece: 7x, 7.1°

χ
ζ2
30 Her
σ
τ

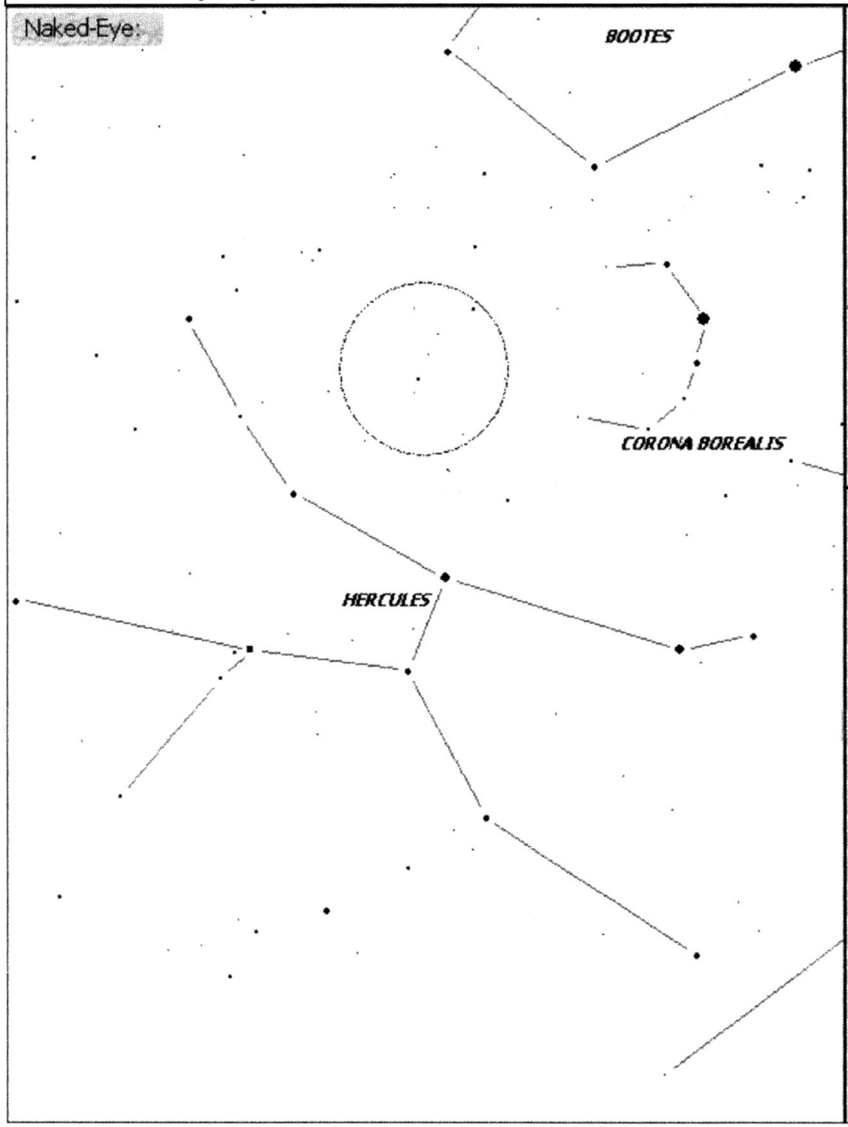

7 x 50 Binoculars

Name	**31 Cygni, Omicron-1, o¹ Cygni**
Location	RA: 20h 13m 38s, Dec: +46° 44' 29" (2000) in Cygnus
Found With	8 x 50 Finder
Description	This remarkable multiple star is one of the best in the sky. The three bright components are easily visible in binoculars and telescopes; each has a distinctive color. Component A is orange- red, C is bluish, and D is white. Some astronomers refer to it as the "Patriotic Star" – the red, white, and blue of the United States flag. 31 Cygni is labeled as Omicron-1 (o¹) Cygni on star charts, component D is labeled as 30 Cygni. The distance between C and D is about 7'.

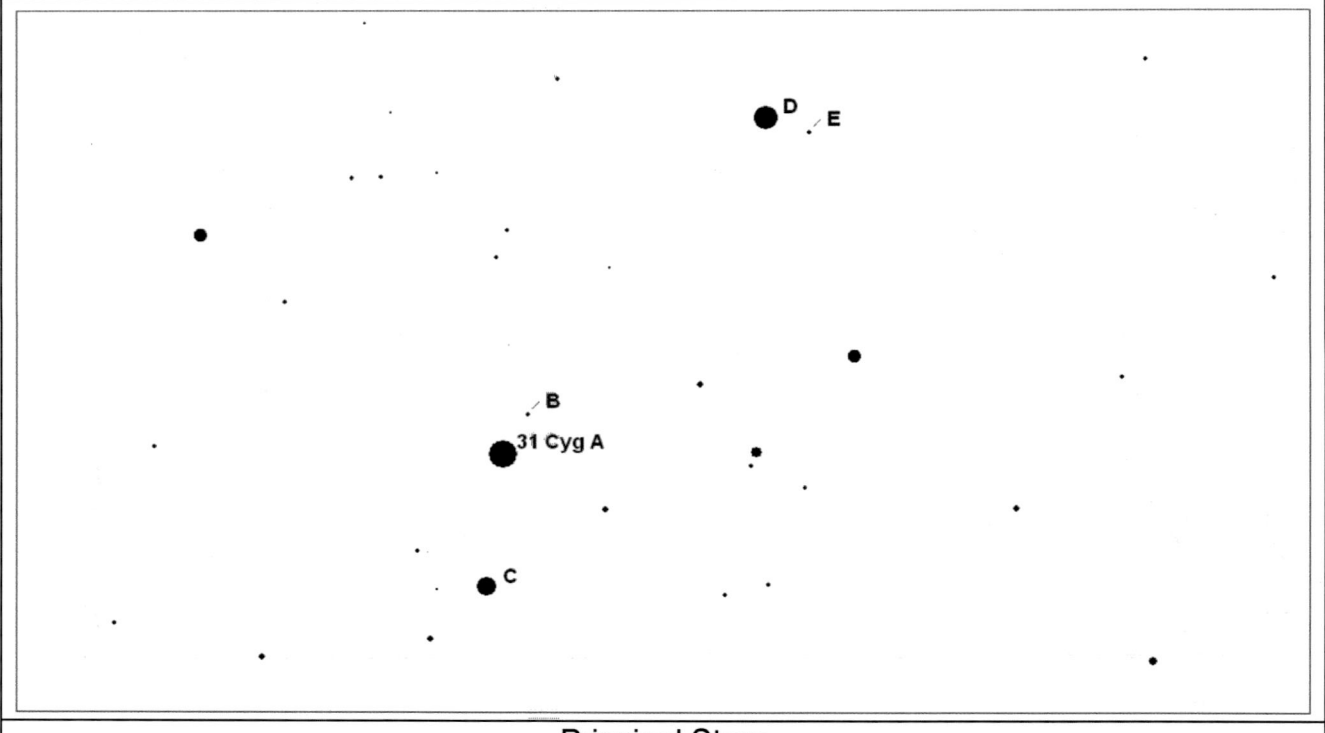

Principal Stars

31 Cyg Multiple Star System
aka V695 Cyg, GCVS 8605, HR 7735, HD 192577, SAO 49337, PPM 59524, HIP 99675
Component A : Mag: V 3.73 to V 3.89 K4
Component C : Mag 6.9 B5V
Component D : Mag 5 A5
Distance: 1400 +/- 320.0 ly

Component	Magnitudes	Other name	Year	PA	Sep
AB:	3.8+13.3 mag	HJ 1495, ADS 13554	In 1878	331°	36.6"
AD	3.8+5 mag	Struve 4050, ADS 13554	In 1835	323°	337.5"
AC	3.8+6.9 mag	Struve 4050, C = HD 192579	In 1836	173°	107"
DE	5+13.2 mag	ES 26, ADS 13554	In 1912	251°	34.3"

Object data from SKYTOOLS™ 2 — used with permission.

31 Cygni, Omicron¹ (o¹) Cygni / Cygnus / RA: 20h 13m 38s, Dec: +46° 44' 29"

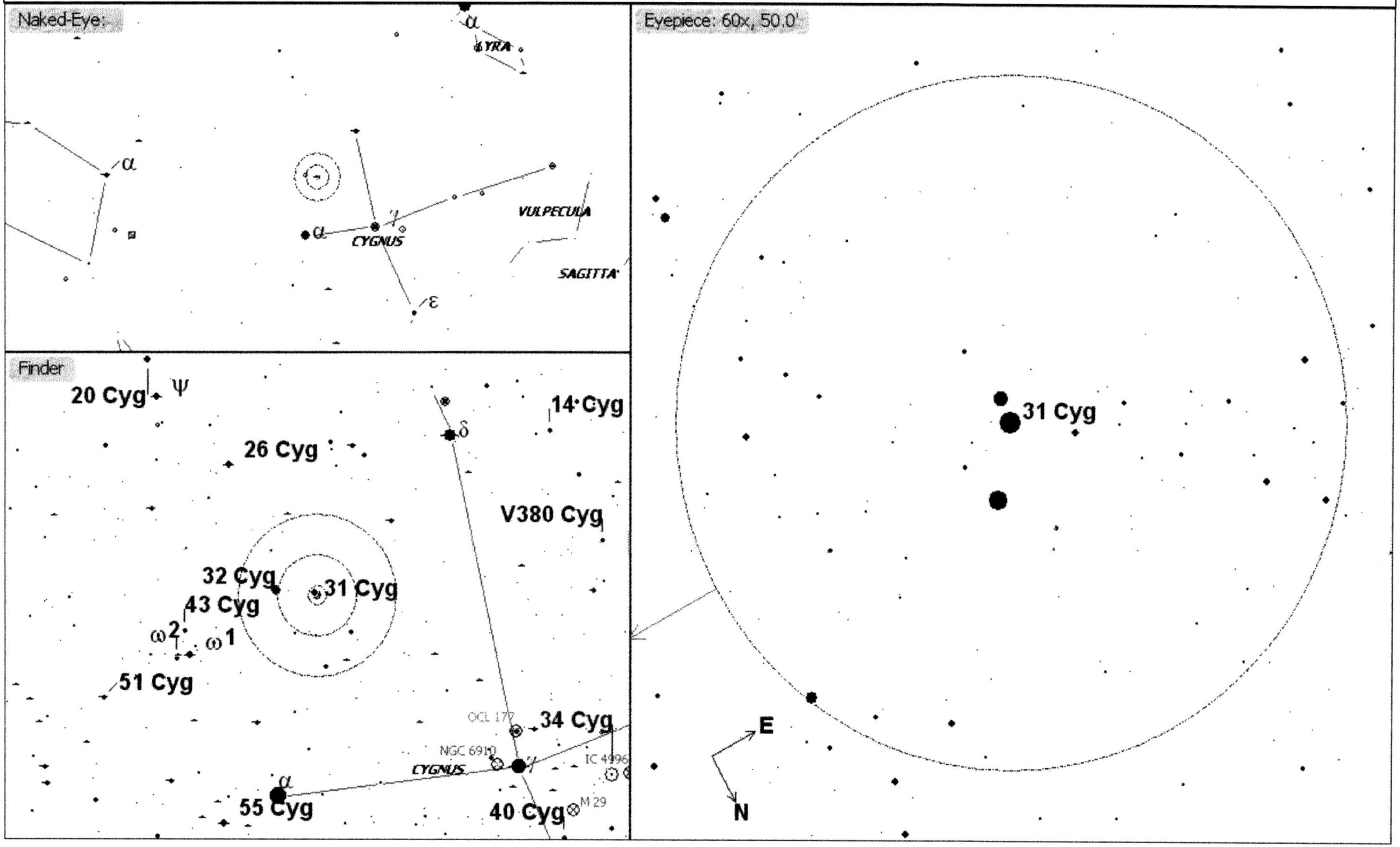

Eyepiece: 60x, 50.0'

Naked-Eye

Finder

20 Cyg ψ
26 Cyg
32 Cyg
43 Cyg ω1
ω2
51 Cyg
14 Cyg
V380 Cyg
δ
31 Cyg
34 Cyg
OCL 177
NGC 6910
CYGNUS
IC 4996
M29
40 Cyg
55 Cyg α

E N

6" f/8 Reflector

85

Name	**The Cross of Draco Asterism**
Location	RA: 18h 21m 33s, Dec: +49° 07' 18" (2000) in Draco
Colorful Stars	SAO 47411, 47417
Found With	8x50 Finderscope
Description	This asterism is very distinctive in the finderscope when seen from a dark-sky location. The distance from SAO 47411 to 47417 is 36'.

Principal Stars

SAO 47402 Single Star
aka PPM 57022, BD +49 02773,
TYC 03533-0280 1
Magnitude: 9.36

SAO 47411 Single Star
aka HD 169221, HR 6886, PPM 57030
Magnitude: 6.39, K0
Distance: 670 +/- 66.0 ly

SAO 47412 Single Star
aka PPM 57034, BD +49 02778,
TYC 03533-0699 1
Magnitude: 8.19

SAO 47413 Single Star
aka PPM 57036, BD +49 02779,
TYC 03533-0395 1
Magnitude: 8.96

SAO 47414 Single Star
aka PPM 57037, BD +49 02780,
TYC 03533-0255 1
Magnitude: 8.67

*SAO 47417 Single Star
aka HD 169305, HR 6891, PPM 57038,
HIP 89981
Magnitude: 5.02, M0
Distance: 750 +/- 82.0 ly

SAO 47427 Single Star
aka HD 169430, PPM 57050, BD +49 02783
Magnitude: 8.67, A2

Object data from SKYTOOLS™ 2 — used with permission.

The Cross of Draco / SAO 47417 / RA: 18h 21m 33s, Dec: +49° 07' 18"

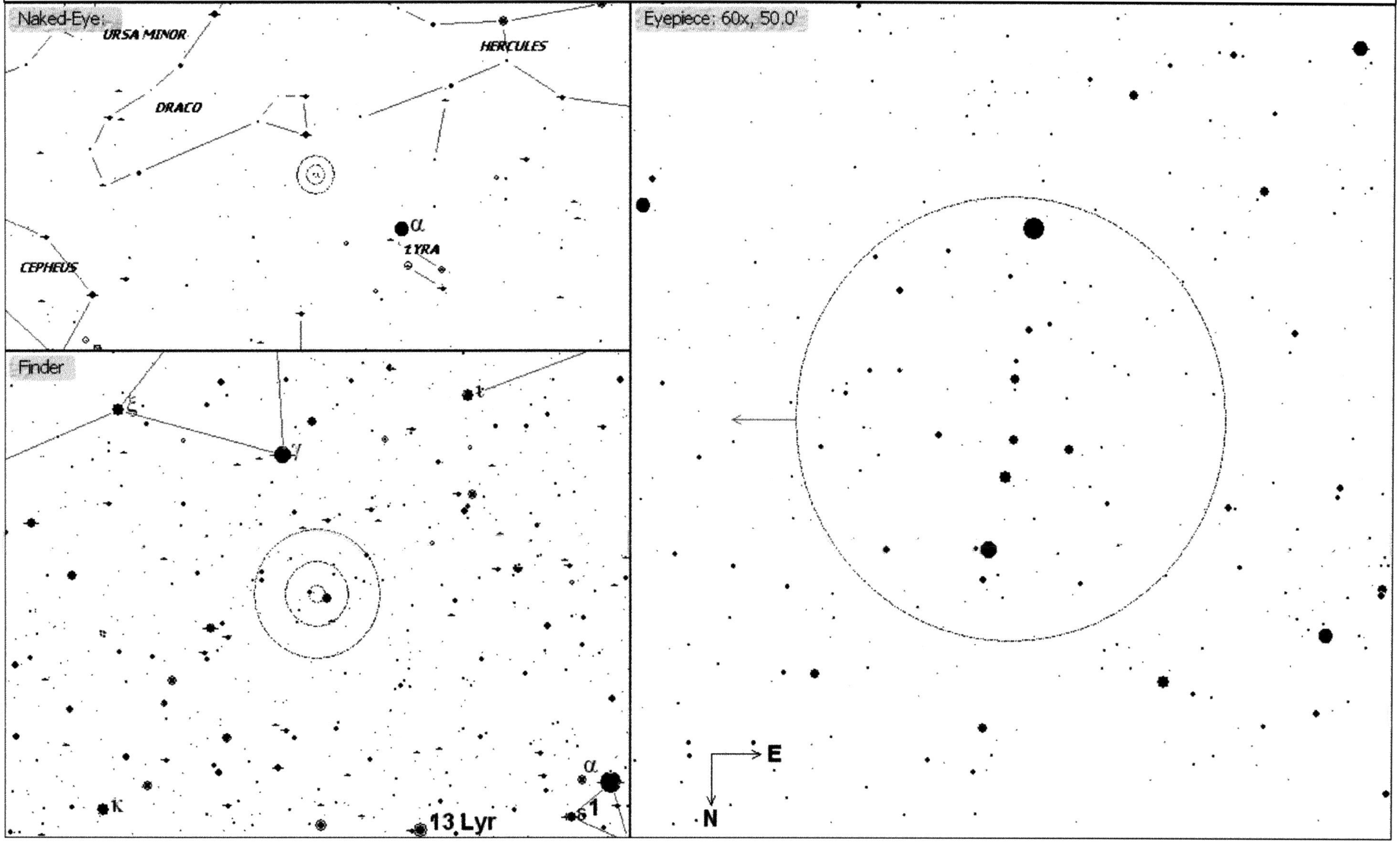

Eyepiece: 60x, 50.0'

Naked-Eye:

URSA MINOR
HERCULES
α
LYRA
DRACO
CEPHEUS

Finder:

α
61
13 Lyr
κ

N E

6" f/8 Reflector

87

Name	**The Kappa (κ) Draconis Asterism**
Location	RA: 12h 33m 29s, Dec: +69° 47' 18" (2000) in Draco
Colorful Stars	4 Dra, 6 Dra, SAO 7611
Found With	Direct Visual Observation
Description	The Kappa Draconis Asterism appears to the naked eye as a fuzzy spot between the Big and Little Dippers. Great color contrast between white Kappa and its orange companions.

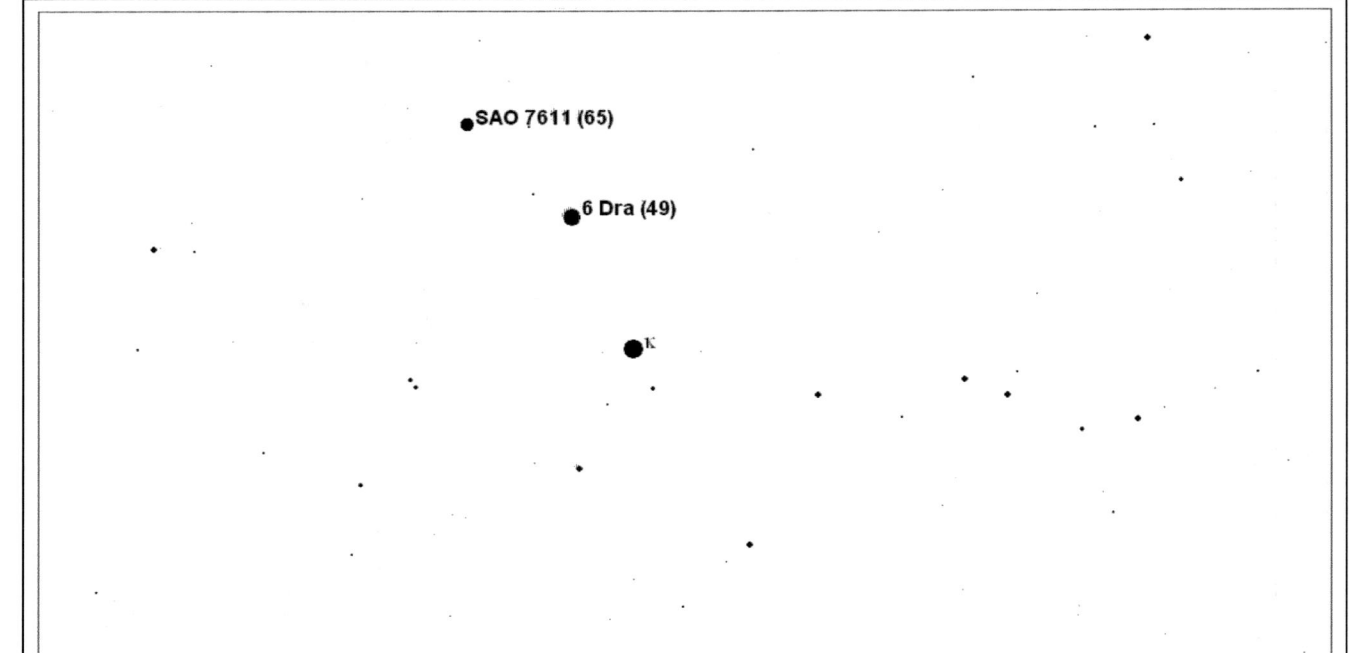

star chart showing SAO 7611 (65), 6 Dra (49), κ, and 4 Dra

Principal Stars

4 Dra Variable Star
aka CQ Dra, GCVS 10120, HR 4765,
HD 108907, SAO 15816
Distance: 580 +/- 53.0 ly
Variability: Type LB
Mag: V 4.95 to V 5.04, M0

6 Dra Single Star
aka HR 4795, HD 109551, SAO 7600
Magnitude: 4.95, K0
Distance: 550 +/- 80.0 ly

SAO 7611 Single Star
aka HD 109822, PPM 8202, HIP 61564
Magnitude: 6.54, K2
Distance: 800 +/- 120.0 ly

*Kappa Dra Variable Star
aka 5 Dra, Kappa Dra, GCVS 10140,
HR 4787, HD 109387, SAO 7593
Distance: 500 +/- 42.0 ly
Variability: Type GCAS
Mag: V 3.82 to V 4.01, B6III

Object data from SKYTOOLS™ 2 — used with permission.

The Kappa Draconis Asterism / Draco / RA: 12h 33m 29s, Dec: +69° 47' 18"

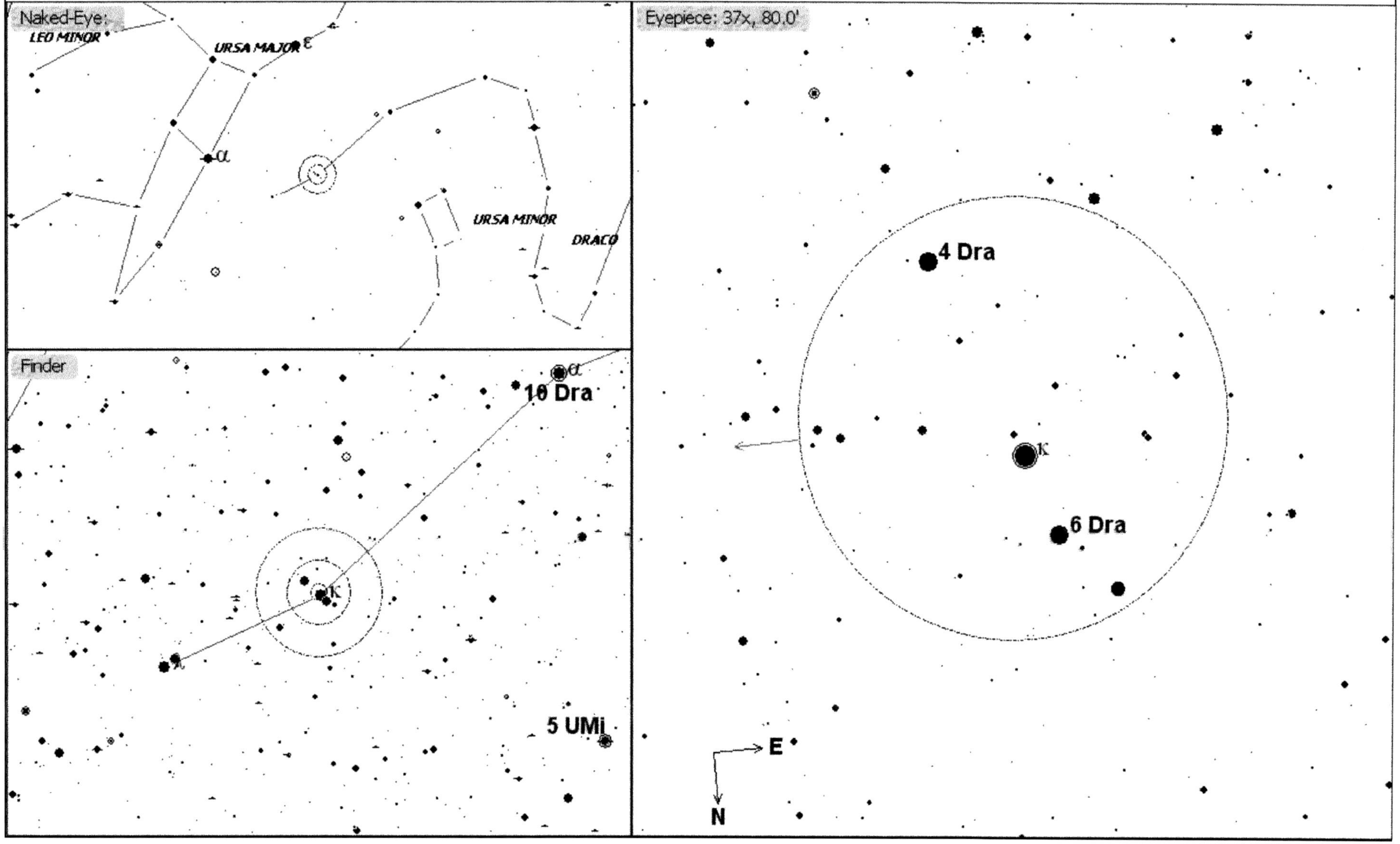

Eyepiece: 37x, 80.0'

4 Dra
6 Dra
κ

E
N

Naked-Eye
LEO MINOR
URSA MAJOR
α
URSA MINOR
DRACO

Finder
α
10 Dra
5 UMi
κ

6" f/8 Reflector

Name	**The Lambda (λ) Draconis Asterism**
Location	RA: 11h 31m 24s, Dec: +69° 19' 52" (2000) in Draco
Colorful Stars	λ Dra, 2 Dra
Found With	Direct Observation
Description	This asterism appears as a fuzzy spot to the naked eye. A telescope reveals a pair of bright orange stars, and an intriguing curved line of fainter stars leading south. This view is a simulation as seen in a reflecting telescope. The distance between Lambda and 2 Dra is 24.5'. The distance between SAO 15331 and 15541 is 25'.

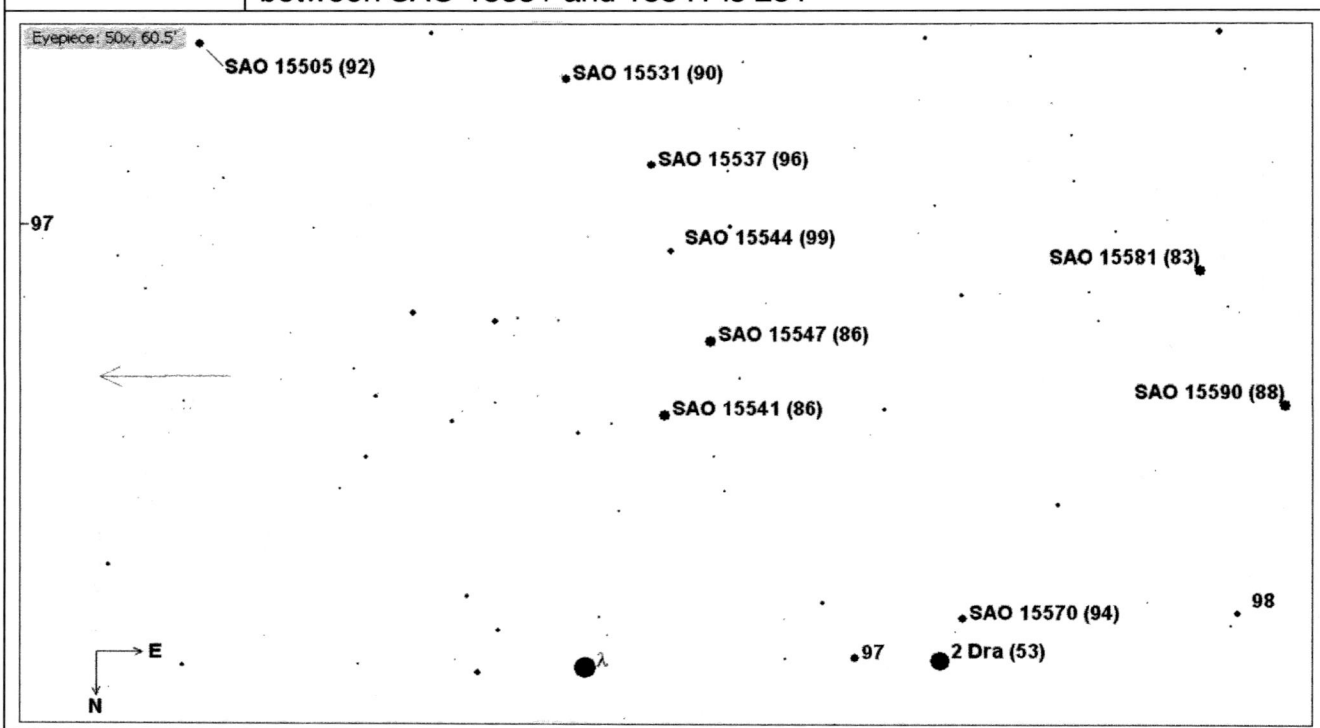

(Finder chart showing the asterism with labeled stars: SAO 15505 (92), SAO 15531 (90), SAO 15537 (96), SAO 15544 (99), SAO 15581 (83), SAO 15547 (86), SAO 15590 (88), SAO 15541 (86), SAO 15570 (94), 97, 98, λ, 2 Dra (53). Eyepiece: 50x, 60.5'. Compass: N down, E right.)

Principal Stars

Visual Components	Telescopic Components : North to South
*Lambda Dra Variable Star aka Giausar, 1 Dra, HR 4434, HD 100029, SAO 15532 Suspected Variable, Type: SR Mag: V 3.78 to V 3.86, M0 Distance: 330 +/- 18.0 ly **2 Dra** Single Star aka HR 4461, HD 100696, SAO 15567 Magnitude: 5.19, K0 Distance: 240 +/- 9.1 ly	**SAO 15541** Single Star aka HD 100176, PPM 17902, HIP 56298 Magnitude: 8.48, G5 Distance: 1000 +/- 270.0 ly **SAO 15547** Single Star aka HD 100253, PPM 17907, HIP 56349 Magnitude: 8.52, G5 Distance: 830 +/- 170.0 ly **SAO 15544** Single Star aka PPM 17903, BD +69 00614, Magnitude: 9.74 **SAO 15537** Single Star aka PPM 17895, BD +69 00612, Magnitude: 9.44 **SAO 15531** Single Star aka HD 99993, PPM 17887, BD +69 00611, GC 15791, TYC 04392-0918 1 Magnitude: 8.91, G5

Object data from SKYTOOLS™ 2 — used with permission.

The Lambda Draconis Asterism / Draco / λ Dra / RA: 11h 31m 24s, Dec: +69° 19' 52"

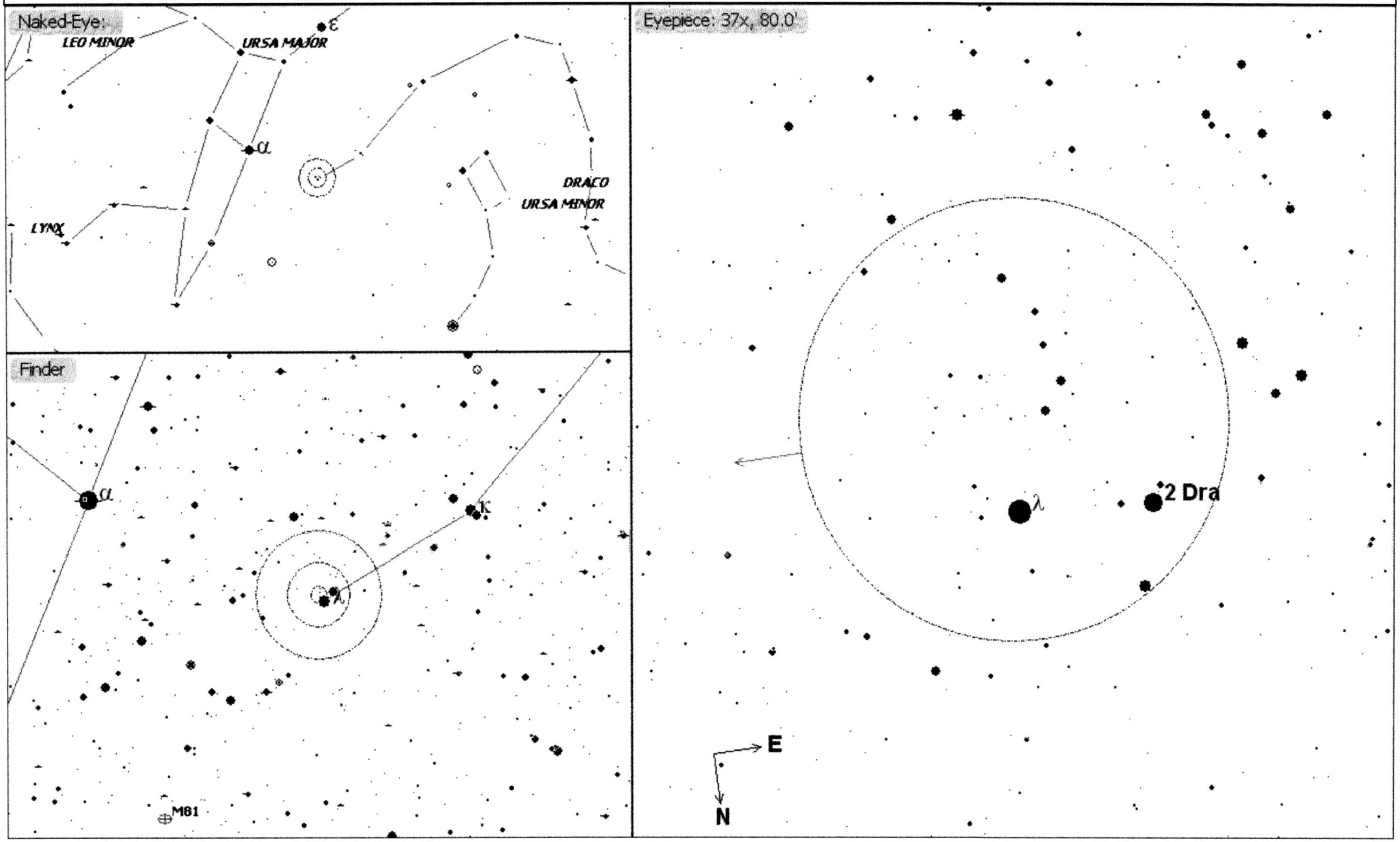

Eyepiece: 37x, 80.0'

λ
·2 Dra

E N

Naked-Eye:
LEO MINOR
URSA MAJOR
ε
α
DRACO
URSA MINOR

Finder:
λ
α
M81

6" f/8 Reflector

Name	**The Draco Pseudo - Cluster**
Location	RA: 12h 03m 36s, Dec: +69° 01' 12" (2000)
Colorful Stars	SAO 15696, 15697, Σ1599 A
Found With	8 x 50 Finderscope
Description	This asterism lies between Kappa and Lambda Draconis, near the faint galaxy NGC 4236. It is a remarkable concentration of stars, 3 of which are the multiples of the famous Struve catalog. The distance between SAO 15686 and Σ3123 is 41'.

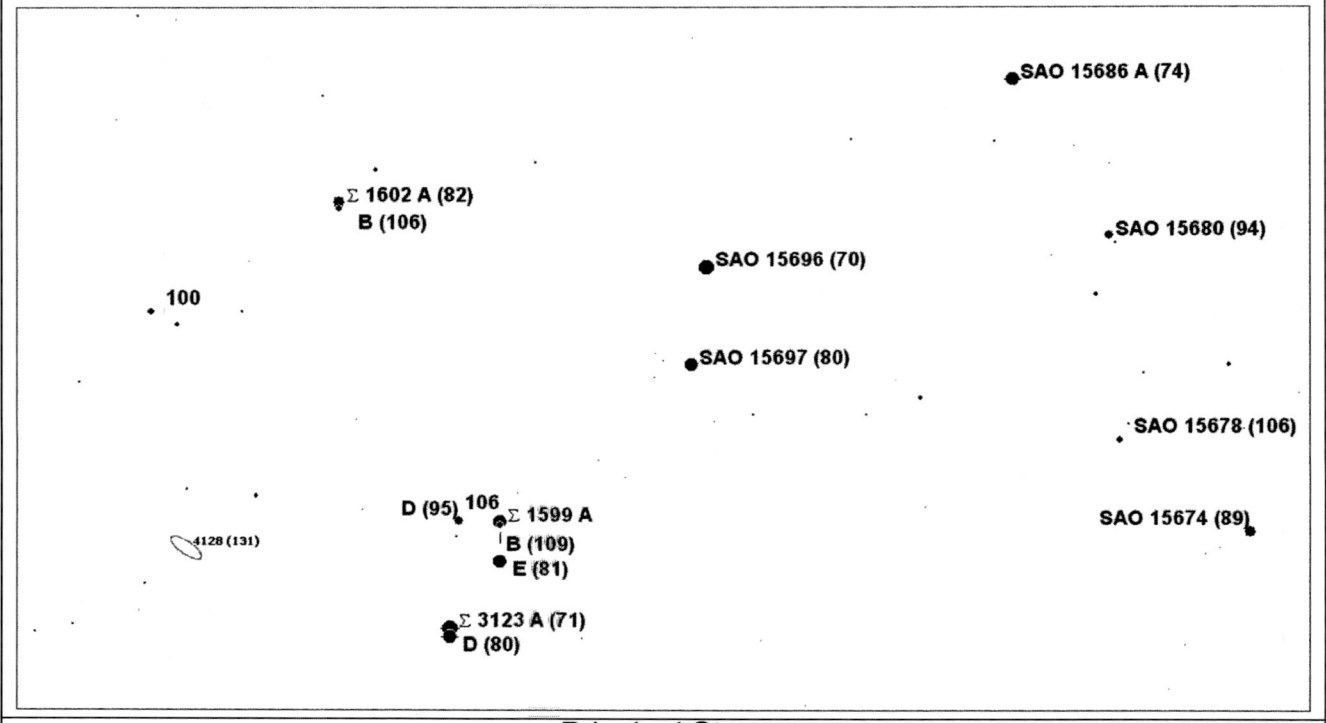

Principal Stars

SAO 15680 Single Star
aka PPM 18076, TYC 04393-0549 1, SI 477779
Magnitude: 9.39

SAO 15686 Multiple Star System
aka HD 104288, PPM 18084, HIP 58567, ADS 8387
Magnitude: 7.42, F0
Distance: 640 +/- 81.0 ly
AB: 7.42+8.81 mag, In 1991: PA 313° Sep 0.12"

*SAO 15696 Single Star
aka HD 104725, PPM 18099, HIP 58798
Magnitude: 7.00, K2
Distance: 630 +/- 74.0 ly

SAO 15697 Single Star
aka HD 104739, PPM 18101, HIP 58808
Magnitude: 7.96, K2
Distance: 720 +/- 110.0 ly

NGC 4128 Galaxy
aka PGC 38555, MCG 12-12-2A, Uppsala 7120
Magnitude: 13.10, Size: 2.3'x 0.8'
Hubble Class: Lenticular, Orientation: Edge on
Position Angle: 58°

NGC 4236 Galaxy (Finder Chart)
Aka PGC 39346, MGC 12-12-4, Uppsala 7306
Magnitude 10.1, Size 22.9' x 7.2'
Hubble Class: barred Spiral

Struve 1599 Multiple Star System
aka SAO 15709, HD 105028, ADS 8413
Magnitude: 7.39, K5
Distance: 800 +/- 200.0 ly
AB: 7.39+10.89 mag; In 1991: PA 166° Sep 10.15"
AC: 7.39+12.6 mag ; In 1911: PA 332° Sep 105.80"
AD: 7.39+9.5 mag, D=SAO 15712;
In 1875: PA 88° Sep 127.00"
AE: 7.39+8.1 mag, E=HD 105029;
In 1911: PA 180° Sep 126.80"

Struve 3123 Multiple Star System
aka HD 105122, SAO 15713, ADS 8419
Magnitude: 7.10, F5
Distance: 430 +/- 39.0 ly
AB: 7.1+8.01 mag, PA 224° Sep 0.25" (2004.4)
AC: 7.1+15.6 mag, In 1895: PA 309° Sep 3.00"
AD: 7.1+8 mag, In 1953: PA 181° Sep 26.00"

Struve 1602 Multiple Star System
aka HD 105287, SAO 15716, ADS 8428
Magnitude: 8.19, G5
Distance: 290 +/- 47.0 ly
AB: 8.19+10.58 mag, B=HIP 59095
In 1991: PA 180° Sep 19.80"

Object data from SKYTOOLS™ 2 — used with permission.

The Draco Pseudo - Cluster / SAO 15696 / RA: 12h 03m 36s, Dec: +69° 01' 12"

Eyepiece: 37x, 80.0'

Naked-Eye

Finder

10 Dra

6" f/8 Reflector

Name	**The Kemble 2 Asterism**
Location	RA: 18h 35m 55s, Dec: +72° 22' 49" (2000) in Draco
Colorful Stars	SAO 9173, 9175, 9176, 9181
Found With	7x50 finderscope
Description	A small asterism caught in the coils of Draco. It resembles the constellation Cassiopeia - almost perfectly. About 1/3 degree separates SAO 9173 and 9189. Named after discoverer Lucian Kemble. This is the classic telescopic asterism, just as Kemble's Cascade is the classic binocular asterism.

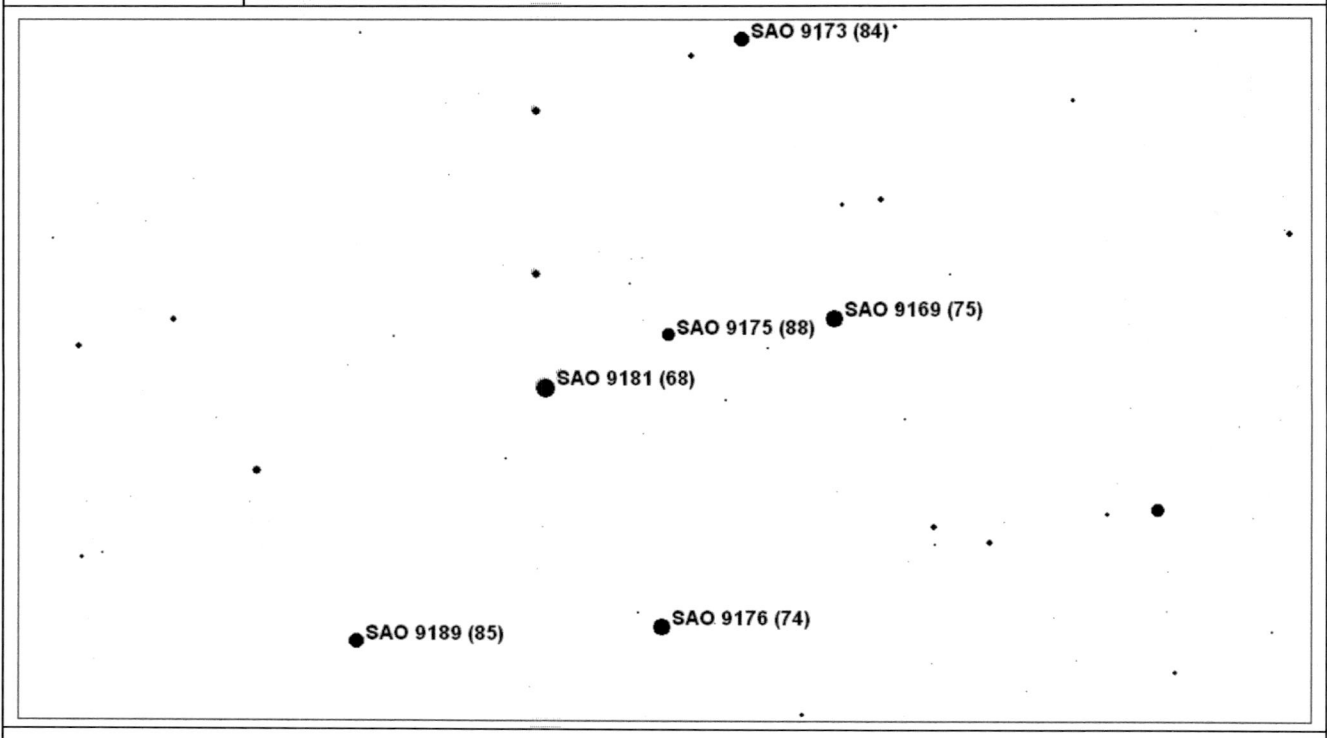

Principal Stars

SAO 9169 Single Star
aka HD 172611, PPM 9913, HIP 91039
Magnitude: 7.50, K0
Distance: 830 +/- 120.0 ly

SAO 9173 Single Star
aka PPM 9915, BD +72 00853
Magnitude: 8.36

SAO 9175 Single Star
aka PPM 9919, BD +72 00854
Magnitude: 8.75

SAO 9176 Single Star
aka HD 172783, PPM 9920, HIP 91121
Magnitude: 7.37, K0
Distance: 530 +/- 48.0 ly

*SAO 9181 Single Star
aka, HD 172922, PPM 9924, HIP 91163
Magnitude: 6.81, K2
Distance: 590 +/- 58.0 ly

SAO 9189 Single Star
aka HD 173127, PPM 9931, BD +72 00858
Magnitude: 8.54, G5

Object data from SKYTOOLS™ 2 — used with permission.

The Kemble 2 Asterism / Draco / SAO 9181 / RA: 18h 35m 55s, Dec: +72° 22' 49"

Eyepiece: 60x, 50.0'

Naked-Eye

Finder

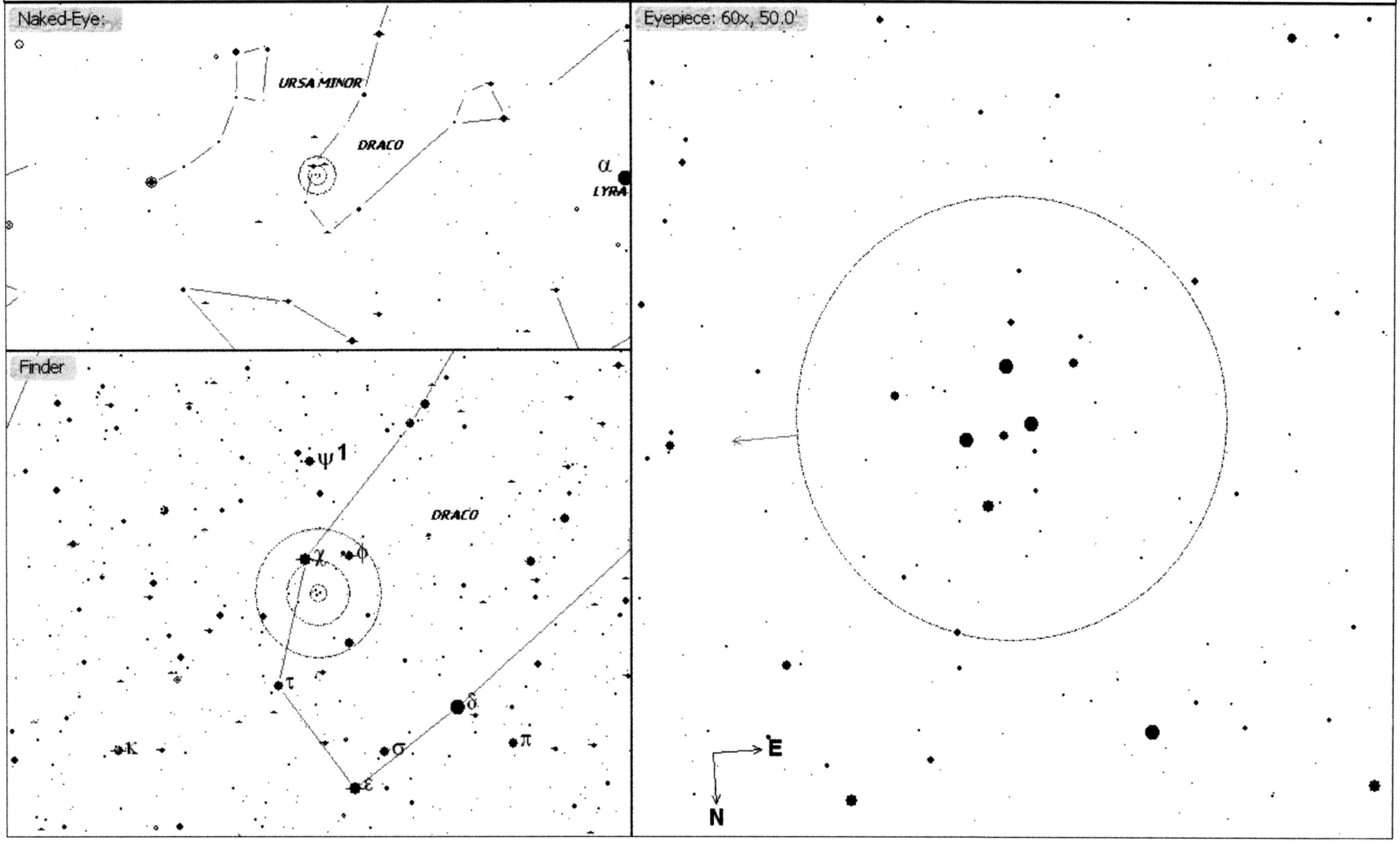

6" f/8 Reflector

Name	**The UX Draconis Asterism**
Location	RA: 19h 21m 36s, Dec: +76° 33' 35" (2000) in Draco
Colorful Stars	UX Dra, SAO 9392, 9365
Found With	7 x 50 finder
Description	A lopsided pentagon made up of (clockwise) UX, SAO 9392, SAO 9354, 59 Dra, and SAO 9365. UX is a red carbon star. With the other stars, this asterism appears as a sparse open cluster. The distance between UX and 59 Dra is 43'.

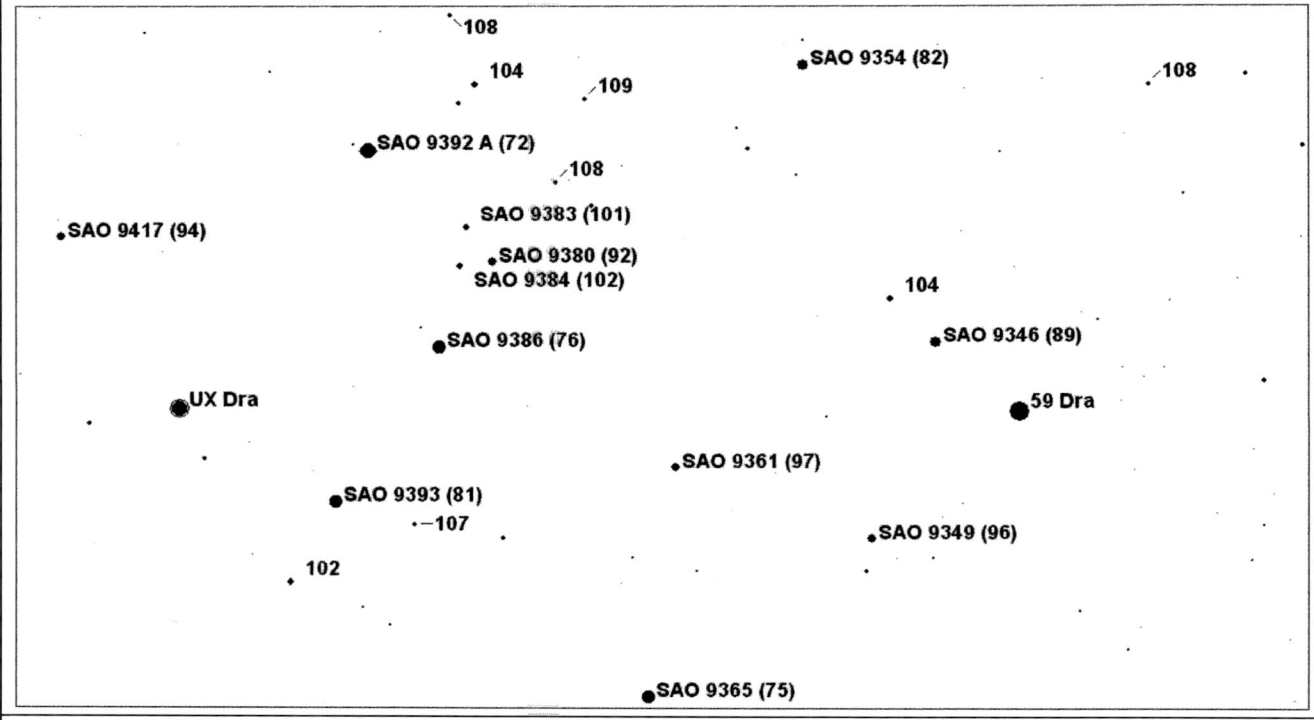

Principal Stars

59 Dra Variable Star
aka HR 7312, HD 180777, SAO 9341
Magnitude: 5.14, A9
Distance: 89 +/- 1.2 ly

SAO 9346 Single Star
aka PPM 10103, BD +76 00719, TYC 04583-2112 1
Magnitude: 8.94

SAO 9349 Single Star
aka PPM 10106, BD +76 00720, TYC 04583-2088 1
Magnitude: 9.60

SAO 9354 Single Star
aka HD 181674, PPM 10109, BD +76 00722
Magnitude: 8.15, A5

SAO 9361 Single Star
aka PPM 10118, BD +76 00724, TYC 04583-2226 1
Magnitude: 9.70

SAO 9365 Single Star
aka HD 182126, PPM 10119, HIP 94554
Magnitude: 7.52, K5
Distance: 1700 +/- 520.0 ly

SAO 9380 Single Star
aka PPM 10137, BD +76 00726, TYC 04583-2001 1
Magnitude: 9.17

SAO 9383 Single Star
aka PPM 10140, BD +76 00728, TYC 04583-1980 1
Magnitude: 10.05

SAO 9384 Single Star
aka PPM 10141, BD +76 00730, TYC 04583-1941 1
Magnitude: 10.15

SAO 9386 Single Star
aka HD 182827, PPM 10143, HIP 94830
Magnitude: 7.58, F0
Distance: 300 +/- 17.0 ly

SAO 9392 Multiple Star System
aka HD 183051, PPM 10151, HIP 94915
Magnitude: 7.23, K0
Distance: 890 +/- 160.0 ly
AB: 7.23+11.02 mag; In 1991: PA 82° Sep 0.75"

SAO 9393 Single Star
aka HD 183076 , PPM 10154, HIP 94945
Magnitude: 8.05, F8
Distance: 260 +/- 15.0 ly

*UX Dra Variable Star
aka GCVS 10040, HD 183556, SAO 9404
Mag: V 5.94 to V 7.10, N0
Distance: 1900 +/- 550.0 ly

Object data from SKYTOOLS™ 2 — used with permission.

The UX Draconis Asterism / Draco / RA: 19h 21m 36s, Dec: +76° 33' 35"

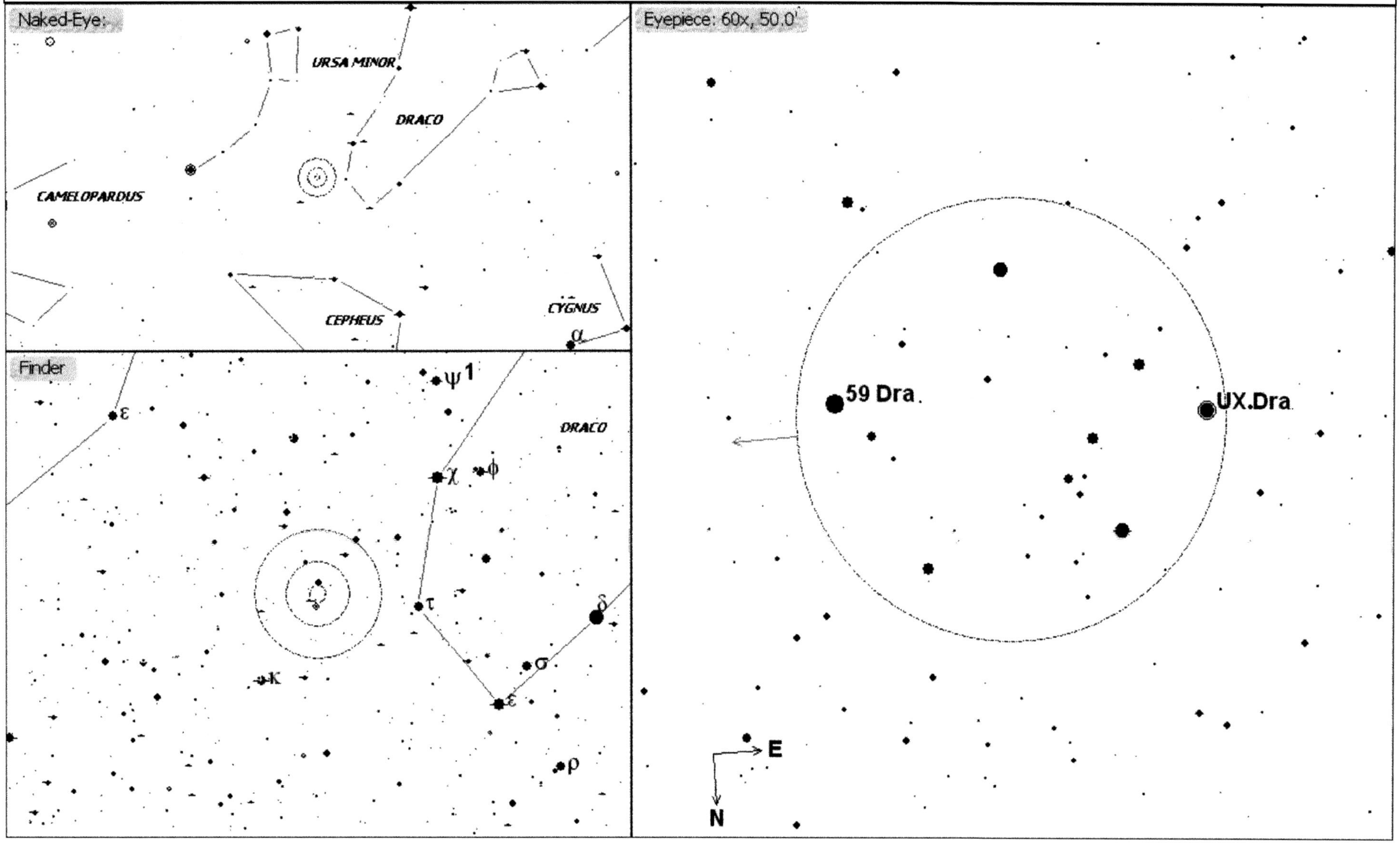

Eyepiece: 60x, 50.0'

Naked-Eye

Finder

6" f/8 Reflector

Name	**The 5-7 Eridani Asterism**
Location	RA: 02h 59m 41s, Dec: -02° 27' 54" (2000) in Eridanus
Colorful Stars	7 Eri
Found With	Direct Observation
Description	This trio of stars appears as a fuzzy object to the naked eye. Amazingly SAO 130215 lacks Bayer or Flamsteed classification. This seems odd considering that fainter stars in the constellation were included. The distance between SAO 130215 and 5 Eri is 24' ; between 5 Eri and 7 Eri is 30' ; between 7 Eri and SAO 130215 is 32'.

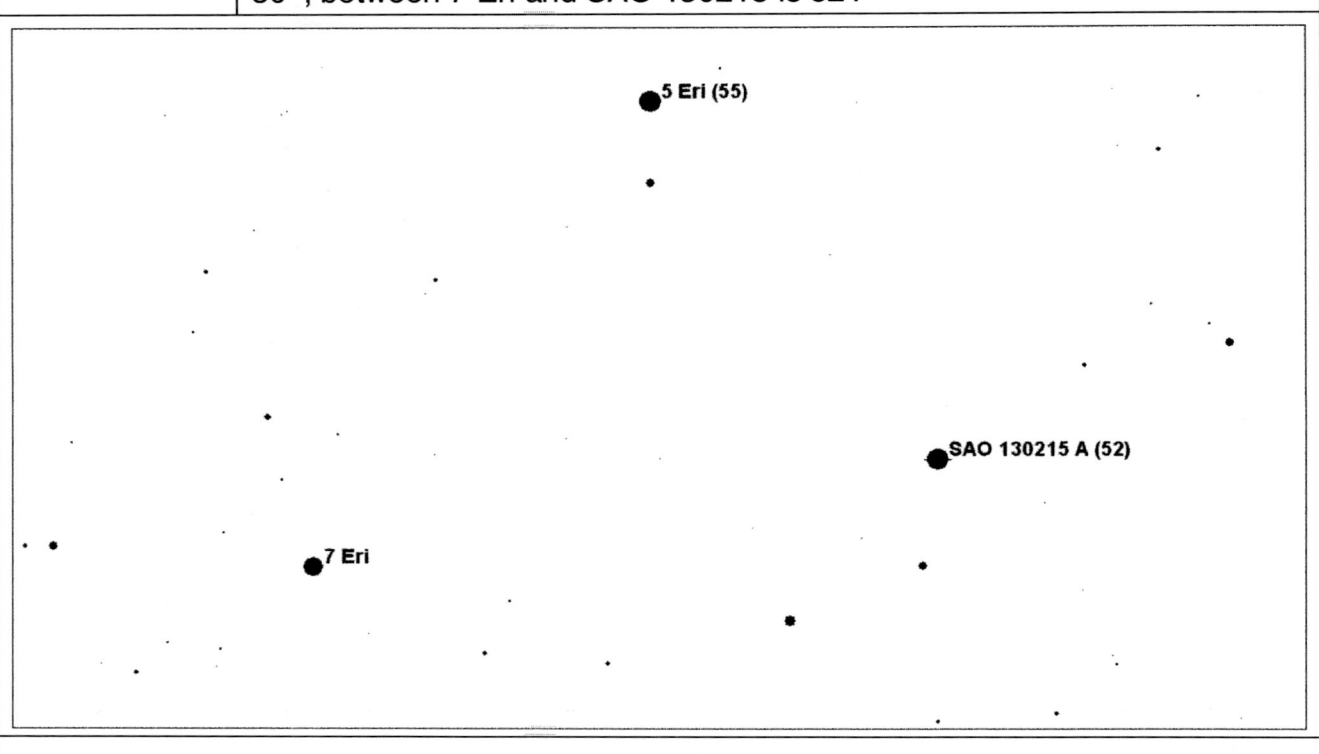

Principal Stars

SAO 130215 Multiple Star System

aka HD 18543, HR 892, PPM 185242

Magnitude: 5.22, A2IV

Distance: 320 +/- 31.0 ly

AB: 5.22+12.5 mag

In 1955: PA 237° Sep 2.30"

*5 Eri Suspected Variable Star

aka HR 899, HD 18633, SAO 130228

V 5.55 Max, Variability Dubious

Mag: H 5.53 to H 5.55,   F5IV

Distance: 330 +/- 35.0 ly

7 Eri   Variable Star

aka CV Eri, GCVS 10287, HR 904,

HD 18760, SAO 130242

Type: LB:

Mag: V 6.10 to V 6.28, M0

Distance: 780 +/- 220.0 ly

Object data from SKYTOOLS™ 2 — used with permission.

The 5-7 Eridani Asterism / Eridanus / 5 Eri / RA: 02h 59m 41s, Dec: -02° 27' 54"

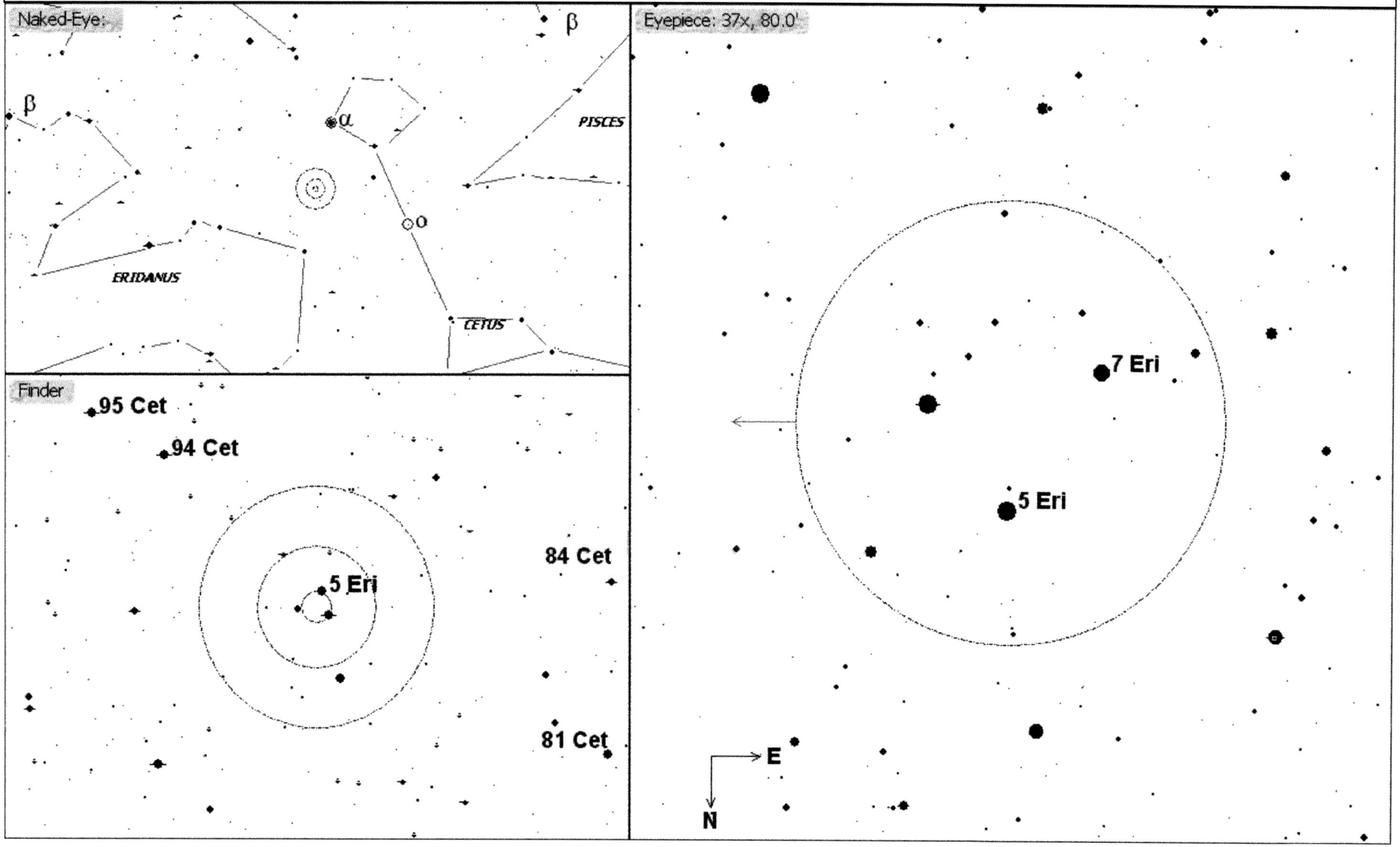

Naked-Eye

Finder

Eyepiece: 37x, 80.0'

7 Eri

5 Eri

E
N

95 Cet
94 Cet
84 Cet
81 Cet
5 Eri

6" f/8 Reflector

Name	**The Rho (ρ) Eridani Asterism**
Location	RA: 03h 02m 42s, Dec: -07° 41' 08" (2000) in Eridanus
Colorful Stars	ρ^1, ρ^2
Found With	Direct Visual Observation
Description	This trio of stars appears as a fuzzy object in Eridanus. The distance between ρ^1 and ρ^3 is 46'

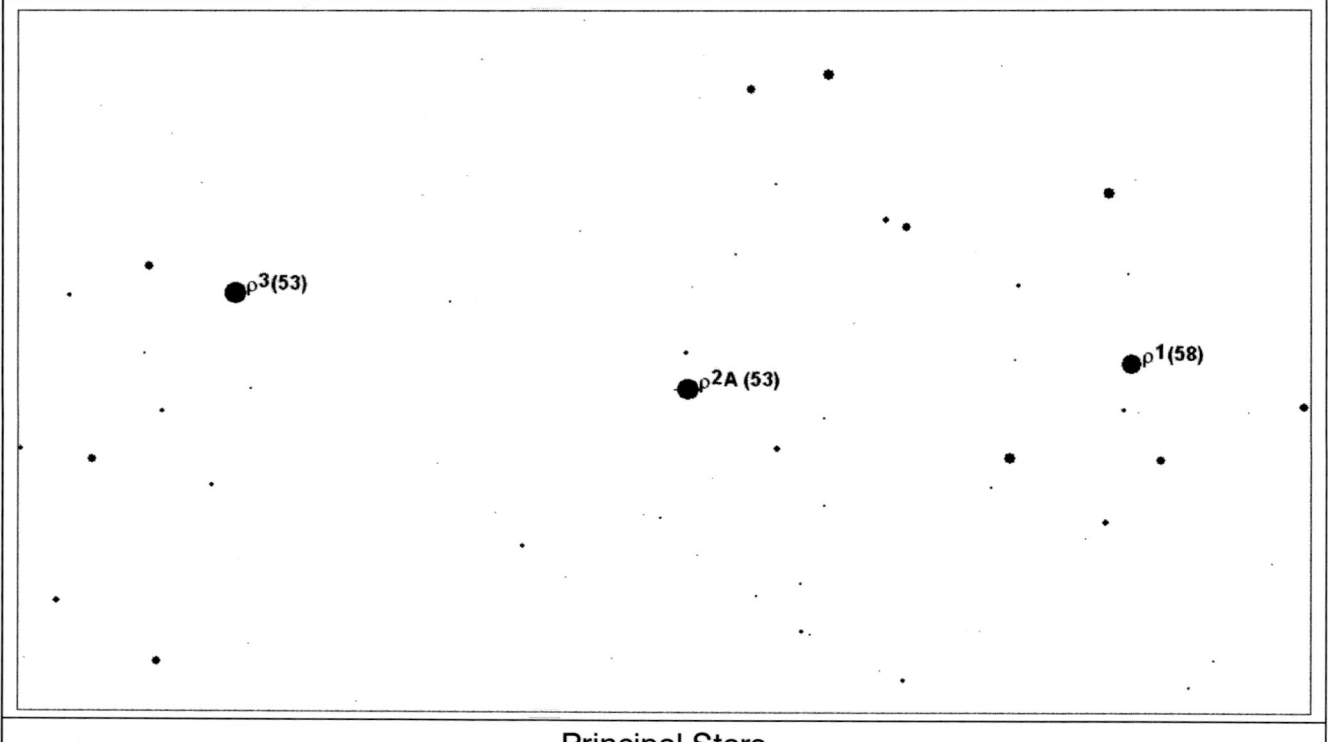

Principal Stars

Rho 1 Eri   Single Star
aka 8 Eri, HR 907, HD 18784, SAO 130243
Magnitude: 5.75, K0II
Distance: 300 +/- 25.0 ly

*Rho 2 Eri   Multiple Star System
aka 9 Eri, HR 917, HD 18953, SAO 130254
BU 11, ADS 2312
Magnitude: 5.32, K0II
Distance: 260 +/- 18.0 ly
AB: 5.32+9.7 mag, B=SAO 130254
In 1959: PA 75° Sep 1.80"

Rho 3 Eri   Single Star
aka 10 Eri, HR 925, HD 19107, SAO 130269
Magnitude: 5.26, A8
Distance: 140 +/- 4.7 ly

Object data from SKYTOOLS™ 2 — used with permission.

The Rho (ρ) Eridani Asterism / Eridanus / ρ² Eri / RA: 03h 02m 42s, Dec: -07° 41' 08"

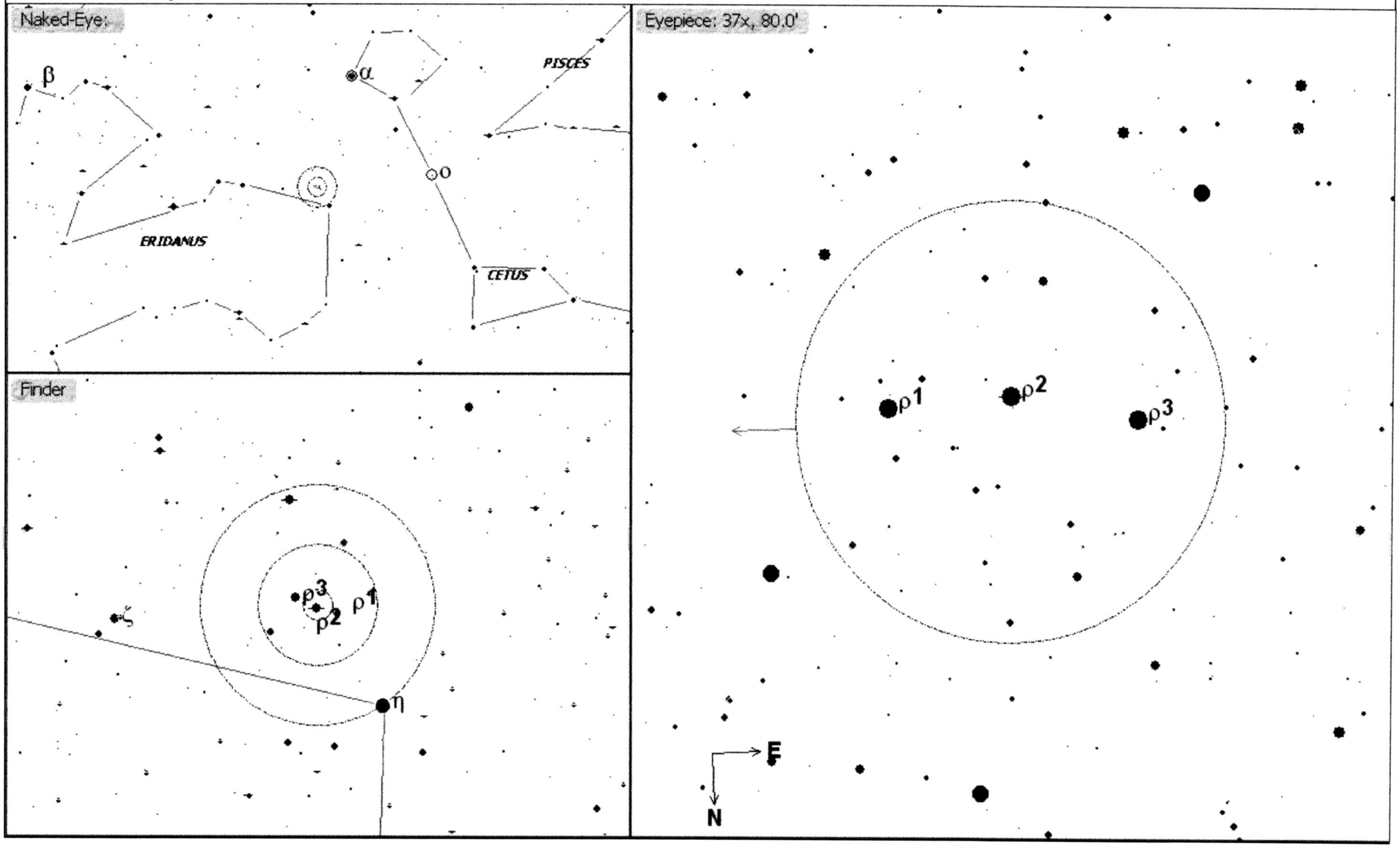

Eyepiece: 37x, 80.0'

Naked-Eye

Finder

6" f/8 Reflector

Name	**The Chi $^{1\text{-}2\text{-}3}$ (χ1χ2χ3) Fornacis Asterism**
Location	RA: 03h 27m 33s, Dec: -35° 40' 53" (2000) in Fornax
Colorful Stars	χ^2
Found With	80mm refractor
Description	The trio of Chi $^{1\text{-}2\text{-}3}$ For is an attractive grouping for small telescopes. Several fainter stars contribute richness to the asterism. The distance from Chi ^{1} to Chi ^{2} is 24'.

```
                                              • SAO 194298 (97)

    • 96                                         • SAO 194299 (94)


                              •χ² (57)

  •SAO 194339 (72)


                         •χ³ A (65)
                •SAO 194327 A (90) •SAO 194315 (93)
                                              •SAO 194308 (89)    •χ¹ •SAO 194288 (99)
                                                                    •SAO 194290 A (73)

                     •SAO 194319 (80)
```

Principal Stars

Chi 1 For Suspected Variable
aka HR 1042, HD 21423,
SAO 194289, NSV 1142
V 6.39 Max, A1IV, Variability Dubious
Distance: 330 +/- 21.0 ly
SAO 194288 Single Star
aka PPM 279020, CD -36 01289
Magnitude: 9.87, F8
SAO 194290 Multiple Star System
aka HD 21434, PPM 279023, B 1449
Magnitude: 7.31, A9
AB: 7.3+8.3 mag
In 1928: PA 216° Sep 0.20"
SAO 194308 Multiple Star System
aka HD 21547, PPM 279048, B 1451
Magnitude: 8.89, F3
AB: 8.86+12.1 mag
In 1928: PA 39° Sep 1.40"

*Chi 2 For Single Star
aka HR 1054, HD 21574, SAO 194312
Magnitude: 5.71, K2III
Distance: 460 +/- 63.0 ly
SAO 194315 Single Star
aka PPM 279064, CD -36 01309,
Magnitude: 9.26, F8
Chi 3 For Multiple Star System
aka HR 1058, HD 21635, SAO 194318, I 58
Magnitude: 6.49, A1
Distance: 360 +/- 34.0 ly
AB: 6.49+10.4 mag, B=HIP 16156
In 1991: PA 247° Sep 6.49"
SAO 194319 Single Star
aka HD 21645, PPM 279069, HIP 16167
Magnitude: 8.03, F2
Distance: 320 +/- 29.0 ly
SAO 194327 Multiple Star System
aka HD 21720, PPM 279079, B 1453
Magnitude: 8.97, F5
AB: 8.99+9.7 mag
In 1942: PA 64° Sep 0.20"

Object data from SKYTOOLS™ 2 — used with permission.

The Chi [1-2-3] ($\chi1\chi2\chi3$) Fornacis Asterism / Fornax / Chi² For / RA: 03h 27m 33s, Dec: -35° 40' 53" (2000)

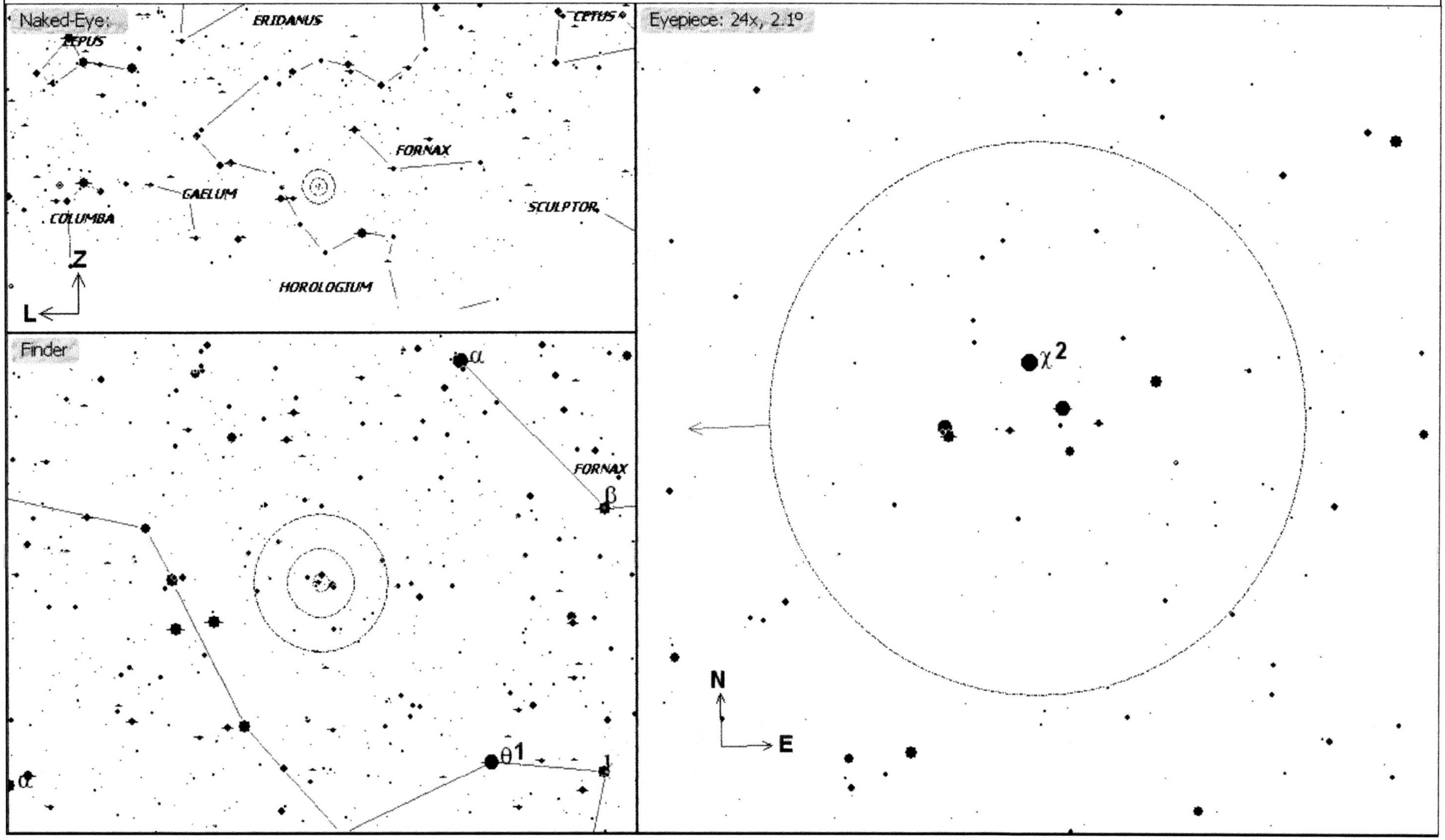

Eyepiece: 24x, 2.1°

Naked-Eye

Finder

80mm f/6 Refractor

Name	**The Chi (χ) Herculis Asterism**
Location	RA: 15 h52 m40s, Dec: +42° 27' 05" (2000) in Hercules
Found With	Direct Vision
Description	Chi, 2 and 4 Herculis appear as a single fuzzy object to the naked eye. Optical aid reveals a scattering of stars around Chi. The distance from Chi to 2 Her is 46 arc -minutes.

```
                        . SAO 45779 (95)
  . SAO 45811 (74)
                          . 2 Her

                                                              . SAO 45748 (88)
                                                            Σ 1982 A (92)
                                           . SAO 45761 (89)
                                         .. SAO 45764 (78)
               . 4 Her (57)           . SAO 45773 (85) . SAO 45757 (74)
                     SAO 45783 (93)
                                          . χ (46)
                                                      SAO 45739 (88)

    . SAO 45805 (85)           . SAO 45771 (84)
```

Principal Stars

SAO 45748   Single Star
aka HD 141847, PPM 54988, HIP 77497
Magnitude: 8.85, K0
Distance: 2800 +/- 2200.0 ly

Struve Σ 1982   Multiple Star System
aka HD 141892, HIP 77530, ADS 9804
Magnitude: 9.24
Distance: 1600 +/- 1900.0 ly
AB: 9.24+10.2 mag, B=HIP 77530
In 1991: PA 299° Sep 4.89"

SAO 45757   Single Star
aka HD 142108, PPM 54998, HIP 77632
Magnitude: 7.39, F5
Distance: 400 +/- 31.0 ly

SAO 45761   Single Star
aka HD 142224, PPM 55004
Magnitude: 8.86, G5

SAO 45764   Single Star
aka HD 142284, PPM 55006, HIP 77706
Magnitude: 7.83, G5
Distance: 760 +/- 130.0 ly

*Chi Her   Single Star
aka 1 Her, HR 5914, HD 142373, SAO 45772
Magnitude: 4.62, G0
Distance: 52 +/- 0.4 ly

SAO 45773   Single Star
aka HD 142394, PPM 55021, HIP 77763
Magnitude: 8.51, A
Distance: 730 +/- 120.0 ly

SAO 45783   Single Star
aka HD 142655, PPM 55029, BD +42 02651
Magnitude: 9.31, G0

2 Her   Variable Star
aka HR 5932, HD 142780, SAO 45788,
NSV 7335
Suspected Variable, Type: UNK
Mag: V 5.33 to V 5.45, F0
Distance: 630 +/- 66.0 ly

4 Her   Single Star
aka HR 5938, HD 142926, SAO 45790,
PPM 55045
Magnitude: 5.73, B8
Distance: 480 +/- 38.0 ly

Object data from SKYTOOLS™ 2 — used with permission.

The Chi (χ) Herculis Asterism / Hercules / Chi Her / RA: 15h 52m 40s, Dec: +42° 27' 05"

Naked-Eye

Eyepiece: 24x, 2.1°

Finder

2 Her

4 Her

N E

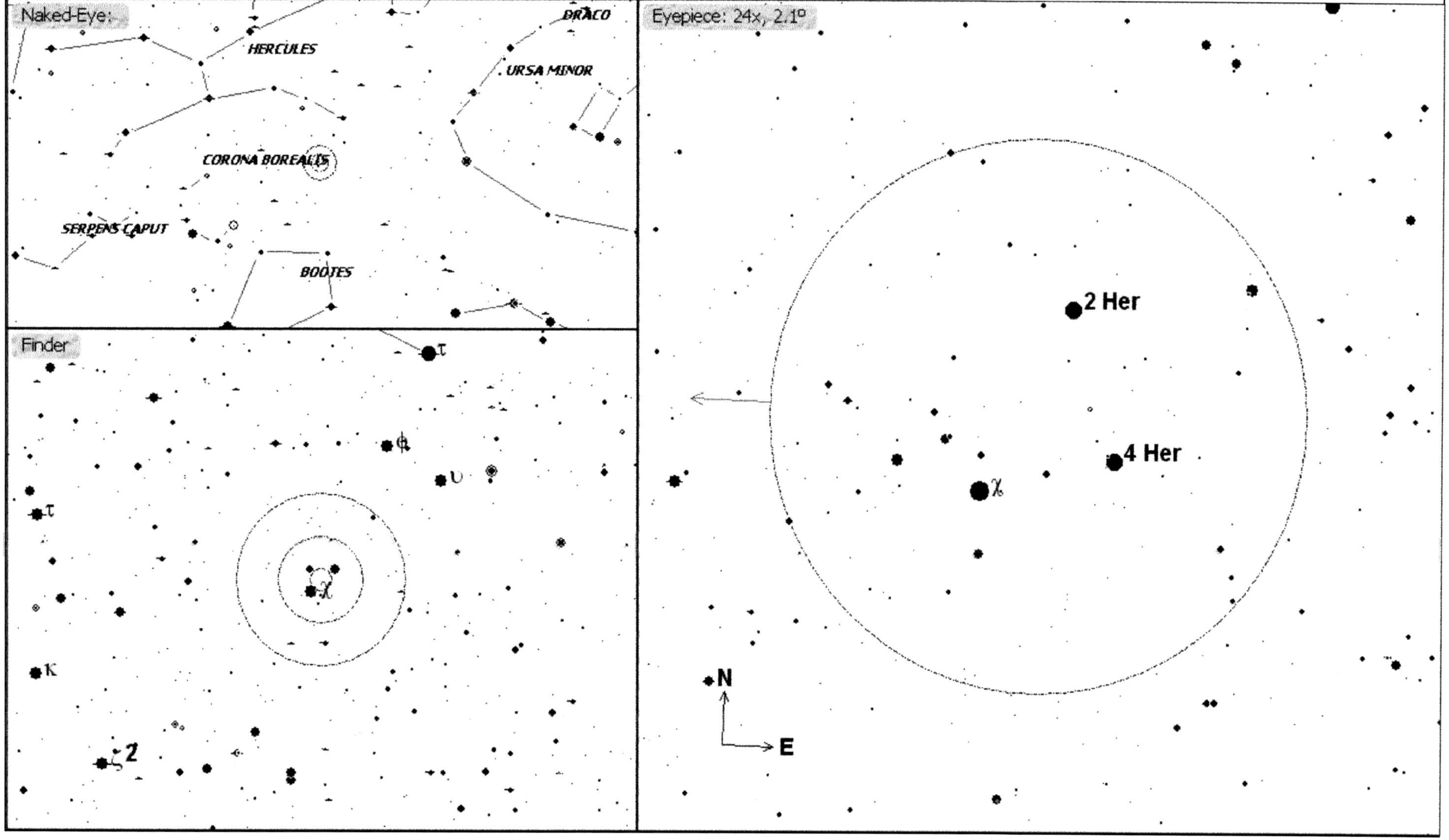

80mm f/6 Refractor

105

Name	**The 60 Herculis Asterism**
Location	RA: 17h 05m 23s, Dec: +12° 44" 27"
Colorful Stars	SAO 102552, 102553
Found With	Direct Visual Observation
Description	This asterism is seen with the naked eye as a fuzzy object about 3 degrees west of Alpha Herculis. It is a line of seven stars from 60 Her to V451 Her, just over 2 degrees in length. The multiple star Struve 4033 is a nice wide pair.

Principal Stars

V451 Her Multiple Star System
aka GCVS 11216, HR 6326, HD 153882, SAO 102536, ADS 10310
Magnitude: V 6.26 to V 6.34 , B9
Distance: 550 +/- 68.0 ly
Variability Type: ACV
Period: 6.009400 days
AB: 6.27+11 mag
In 1783: PA 239° Sep 19.00"

SAO 102552 Single Star
aka HD 154101, PPM 132945, HIP 83411
Magnitude: 7.72, K5
Distance: 1300 +/- 510.0 ly

SAO 102554 Single Star
aka HD 154160, HR 6339, PPM 132947, HIP 83435
Magnitude: 6.52, K0
Distance: 120 +/- 3.4 ly

SAO 102553 Variable Star
aka HD 154143,HR 6337 , PPM 132946
Magnitude: V 4.96 to V 5.05 , M3III
Distance: 410 +/- 35.0 ly

Struve 4033 Multiple Star System
aka HD 154228, HR 6341, SAO 102564
Magnitude: 5.91, A1
Distance: 260 +/- 15.0 ly
AB: 5.91+6.1 mag, B=HR 6342
In 1923: PA 116° Sep 299.20"

*60 Her Multiple Star System
aka HR 6355, HD 154494, SAO 102584
Magnitude: 4.91, A3IV
Distance: 140 +/- 4.6 ly

Object data from SKYTOOLS™ 2 — used with permission.

The 60 Herculis Asterism / Hercules / RA: 17h 05m 23s, Dec: +12° 44' 27"

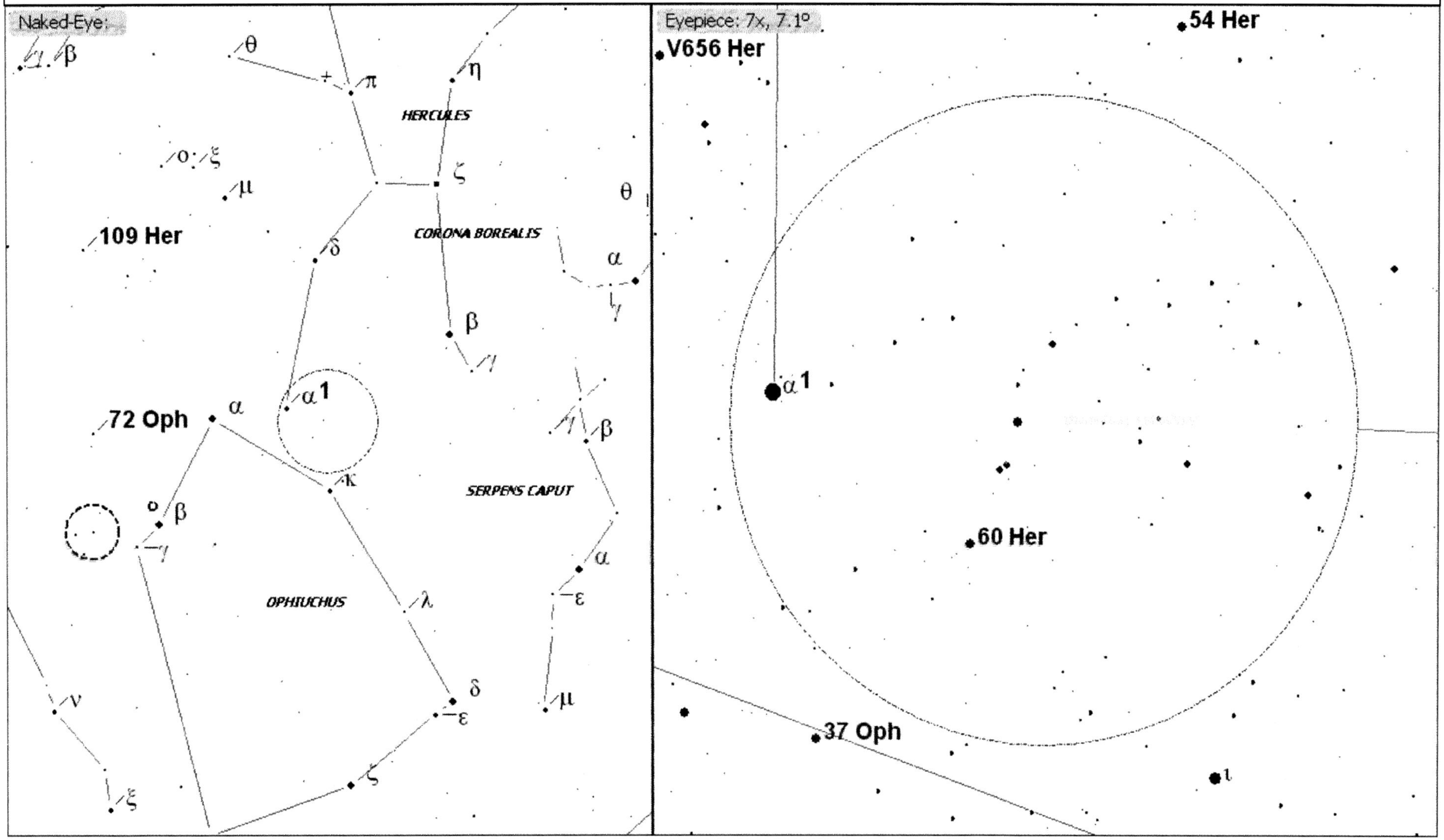

Naked-Eye:

Eyepiece: 7x, 7.1°

- 54 Her
- V656 Her
- α1
- 60 Her
- 37 Oph

7 x 50 Binoculars

Name	**The 1-C-2 Hydrae Asterism**
Location	RA: 08h 25m 40s, Dec −3° 54" 23" in Hydra
Found With	Direct Visual Observation
Description	This trio appears as a fuzzy spot to the naked eye. Optical aid reveals a half-degree long chain of three stars, with 2 Hya being an easy double. SAO 135896 is labeled "C Hya" in most modern star charts. In others it also carries the Flamsteed designation of 30 Mon! A truly interesting history is behind this bright, little known star.

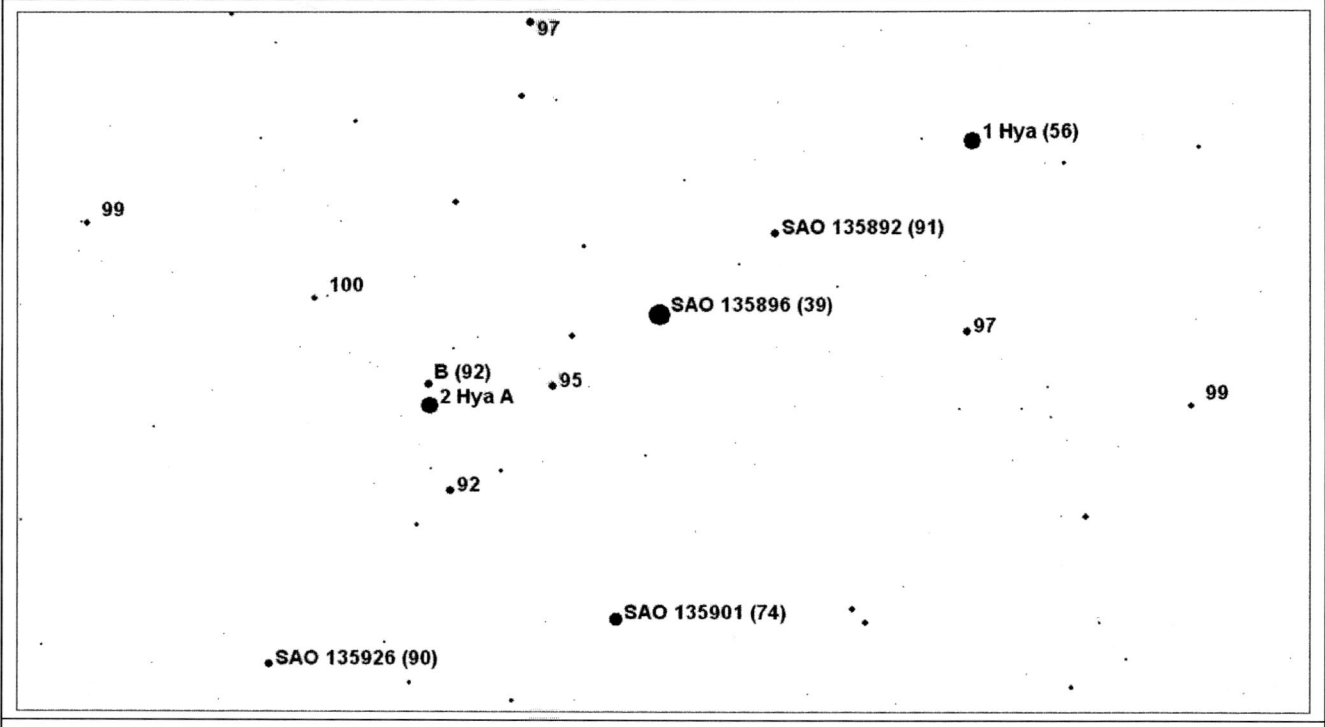

Principal Stars	
1 Hya Single Star aka HR 3297, HD 70958, SAO 135877 Magnitude: 5.61, G2 Distance: 89 +/- 2.1 ly ***SAO 135896** Single Star aka C Hya, 30 Mon, HR 3314, HD 71155 Magnitude: 3.90, A0 Distance: 120 +/- 3.7 ly	**2 Hya** Multiple Star System aka HR 3321, HD 71297, SAO 135916 Magnitude: 5.60, A5 Distance: 170 +/- 7.4 ly AB: 5.6+9.2 mag, In 1901: PA 3° Sep 72.8"

Object data from SKYTOOLS™ 2 — used with permission.

The 1-C-2 Hydrae Asterism / Hydra / C Hya / SAO 135896 / RA: 08h 25m 40s, Dec: -03° 54'23"

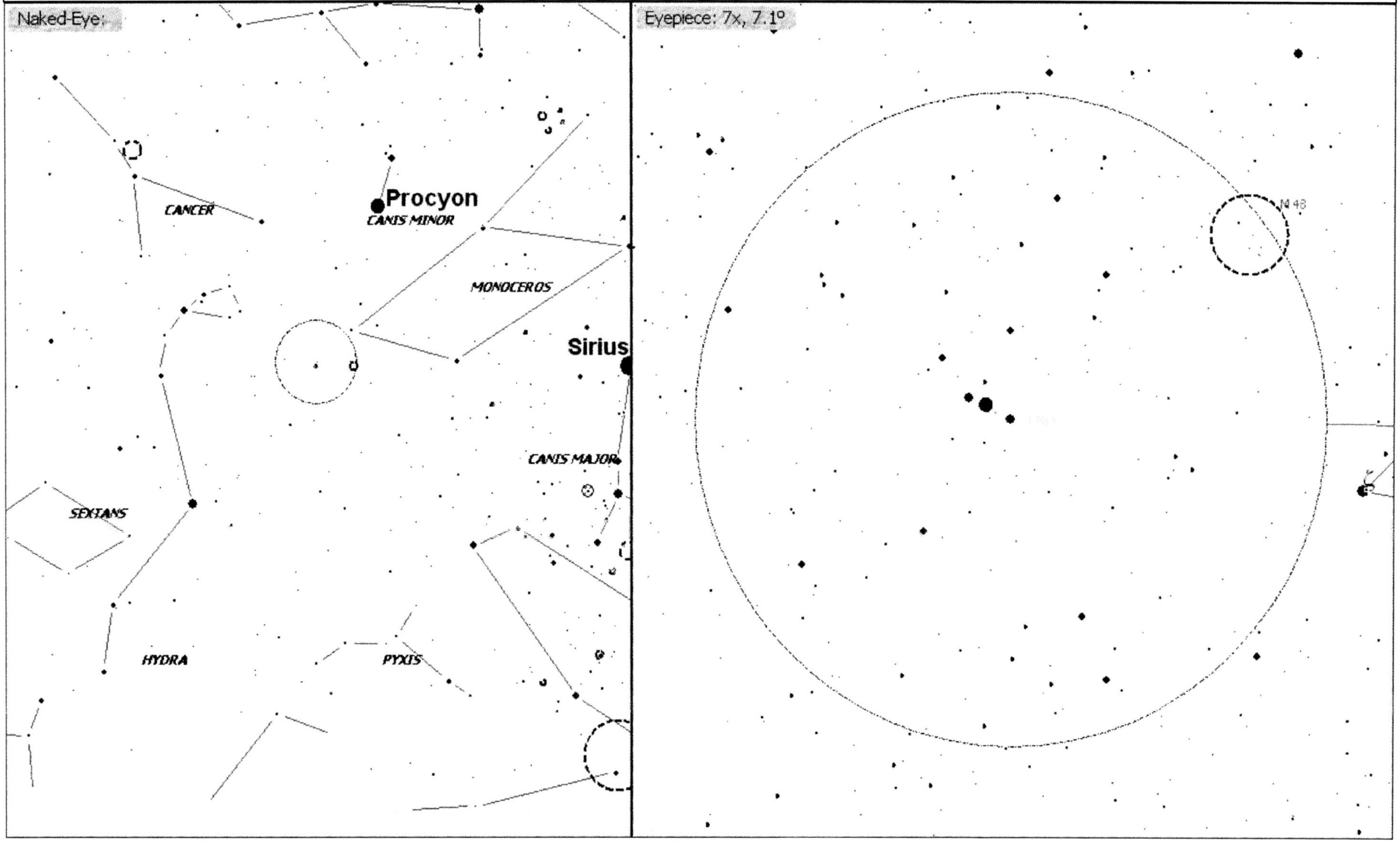

Eyepiece: 7x, 7.1°

Naked-Eye

7 x 50 Binoculars

109

Name	**The "I" Hydrae Asterism**
Location	RA: 09h 41m 17s, Dec: -23° 35' 29" (2000) in Hydra
Found With	Direct Visual Observation
Description	This is the only conspicuous visual fuzzy object south of Alphard in the spring sky. It is a pair for moderately faint stars for the naked eye. SAO 177840 is labeled on star charts as I Hydrae (Roman letter "I") The separation between the pair is 23 arc min.

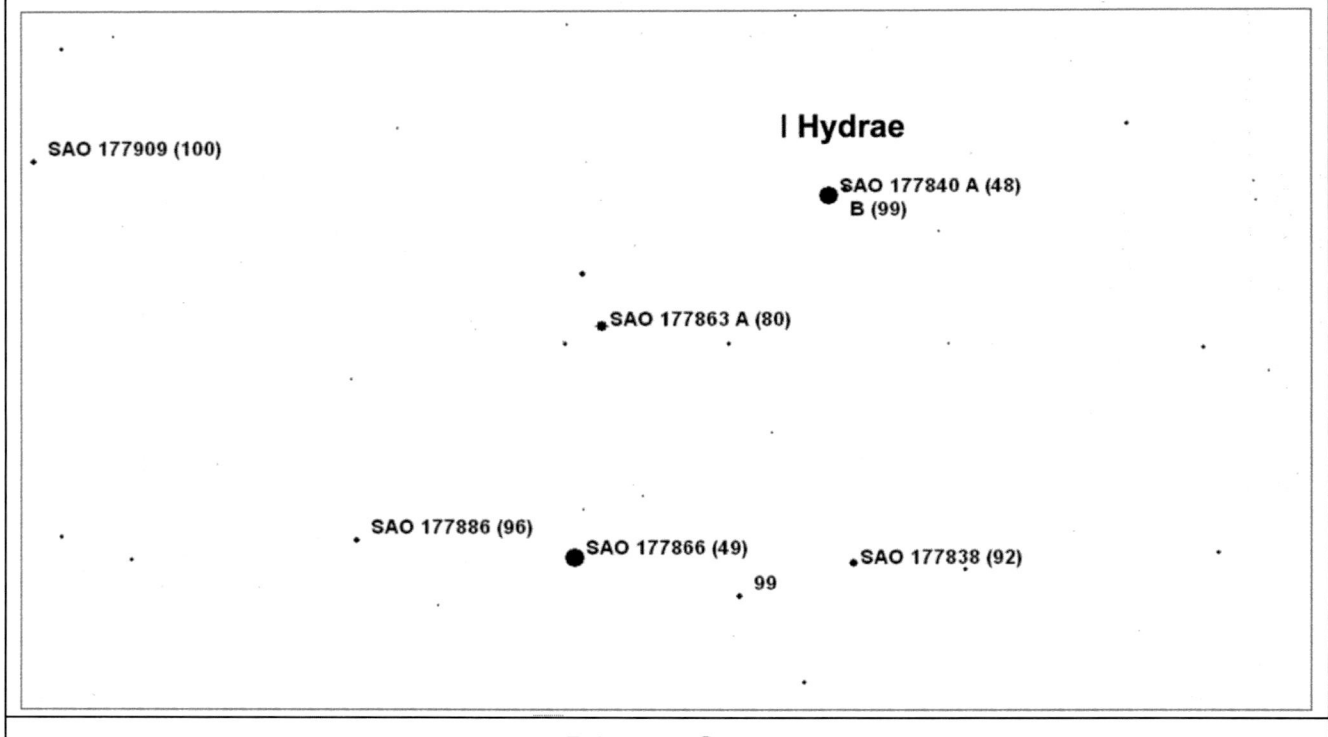

[Star chart showing I Hydrae field with labeled stars: SAO 177909 (100), SAO 177840 A (48) B (99), SAO 177863 A (80), SAO 177886 (96), SAO 177866 (49), SAO 177838 (92), 99]

Principal Stars	
*SAO 177840 Multiple Star System Aka I Hya, HD 83953, HR 3858, H 20 Unsolved variable. Mag: H 4.71 to H 4.74, B5 Distance: 500 +/- 49.0 ly AB: 4.76+9.9 mag In 1904: PA 292° Sep 54.70"	**SAO 177866** Single Star aka HD 84117, HR 3862, PPM 256602 Magnitude: 4.91, F9IV Distance: 49 +/- 0.5 ly

Object data from SKYTOOLS™ 2 — used with permission.

The I Hydrae (Roman Letter I) Asterism / Hydra / RA: 09h 41m 17s, Dec: -23° 35' 29"

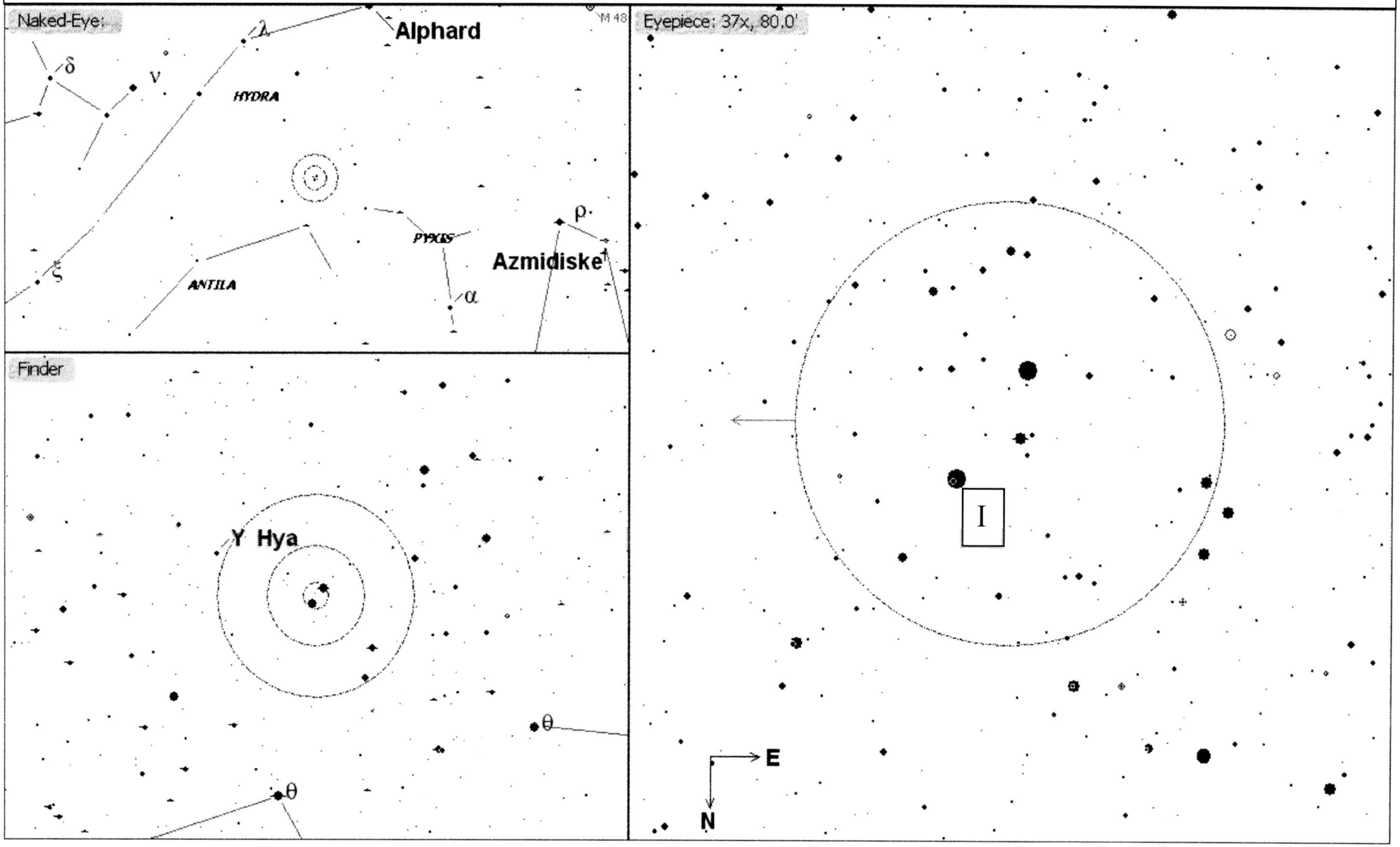

Eyepiece: 37x, 80.0

Naked-Eye

Finder

6" f/8 Reflector

Name	**The Little Cassiopeia Asterism**
Location	RA 22h 31m 17s, Dec.: +50° 16' 57" in Lacerta
Colorful Stars	Beta Lac, 5 Lac
Found With	Direct Vision
Description	The five brightest stars of Lacerta resemble the constellation Cassiopeia. From Beta Lac to 2 Lac is 5.75 degrees. Measuring between the corresponding stars of Cassiopeia, Beta Cas to Epsilon Cas, is 13.25 degrees. See also: Kemble 2 in Draco.

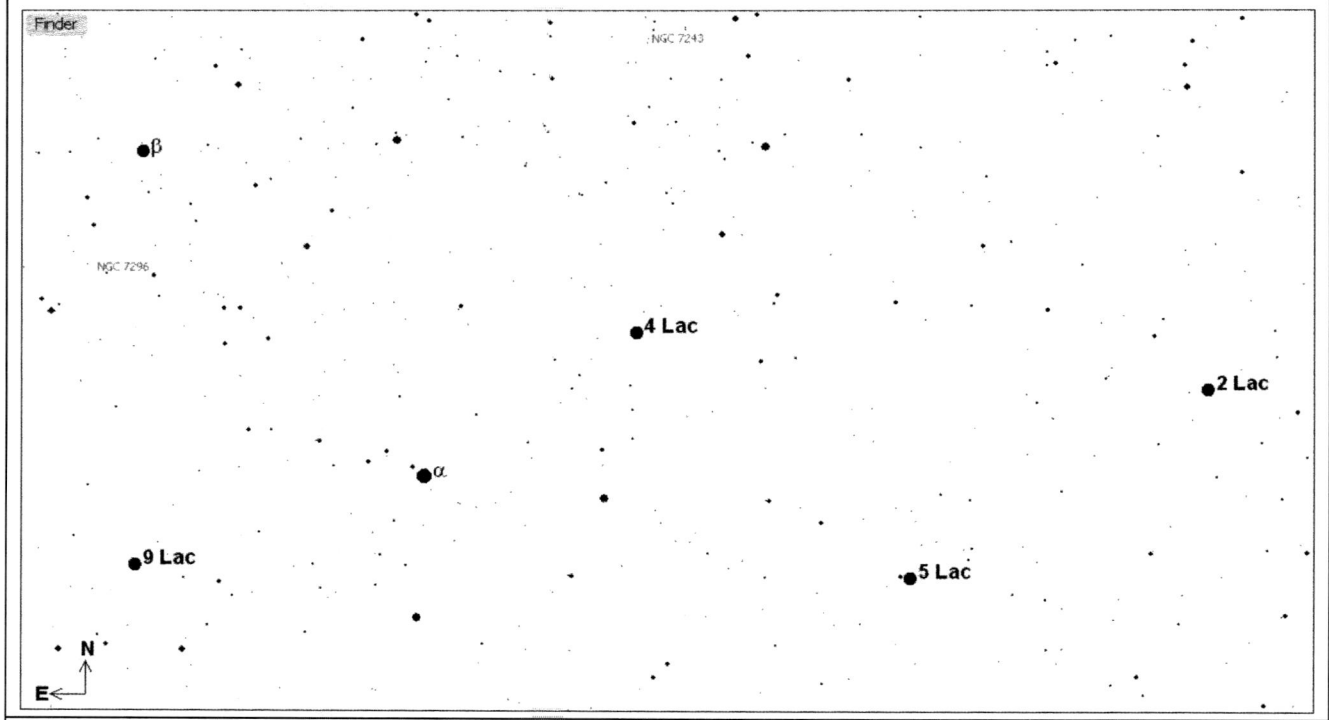

Principal Stars

Beta Lac Single Star
aka 3 Lac, HR 8538, HD 212496, SAO 34395
Magnitude: 4.44, K0
Distance: 170 +/- 4.5 ly

*Alpha Lac Multiple Star System
aka 7 Lac, HR 8585, HD 213558, SAO 34542, BU 703, ADS 16021
Magnitude: 3.77, A1
Distance: 100 +/- 1.7 ly
AB: 3.76+11.8 mag
In 1925: PA 294° Sep 36.30"

4 Lac Single Star
aka HR 8541, HD 212593, SAO 51970
Magnitude: 4.58, B9I
Distance: 2100 +/- 710.0 ly

5 Lac Variable Star
aka HR 8572, HD 213310, SAO 52055
Magnitude: 4.34, M0
Distance: 1200 +/- 210.0 ly

2 Lac Multiple Star System
aka HR 8523, HD 212120, SAO 51904, ADS 15862, HJ 1755
Magnitude: 4.56, B6
Distance: 510 +/- 41.0 ly
AB: 4.55+10.9 mag,
In 1879: PA 9° Sep 48.20"

Object data from SKYTOOLS™ 2 — used with permission.

The Little Cassiopeia Asterism / Lacerta / Alpha Lacertae / RA: 22h 31m 17s, Dec: +50° 16' 57"

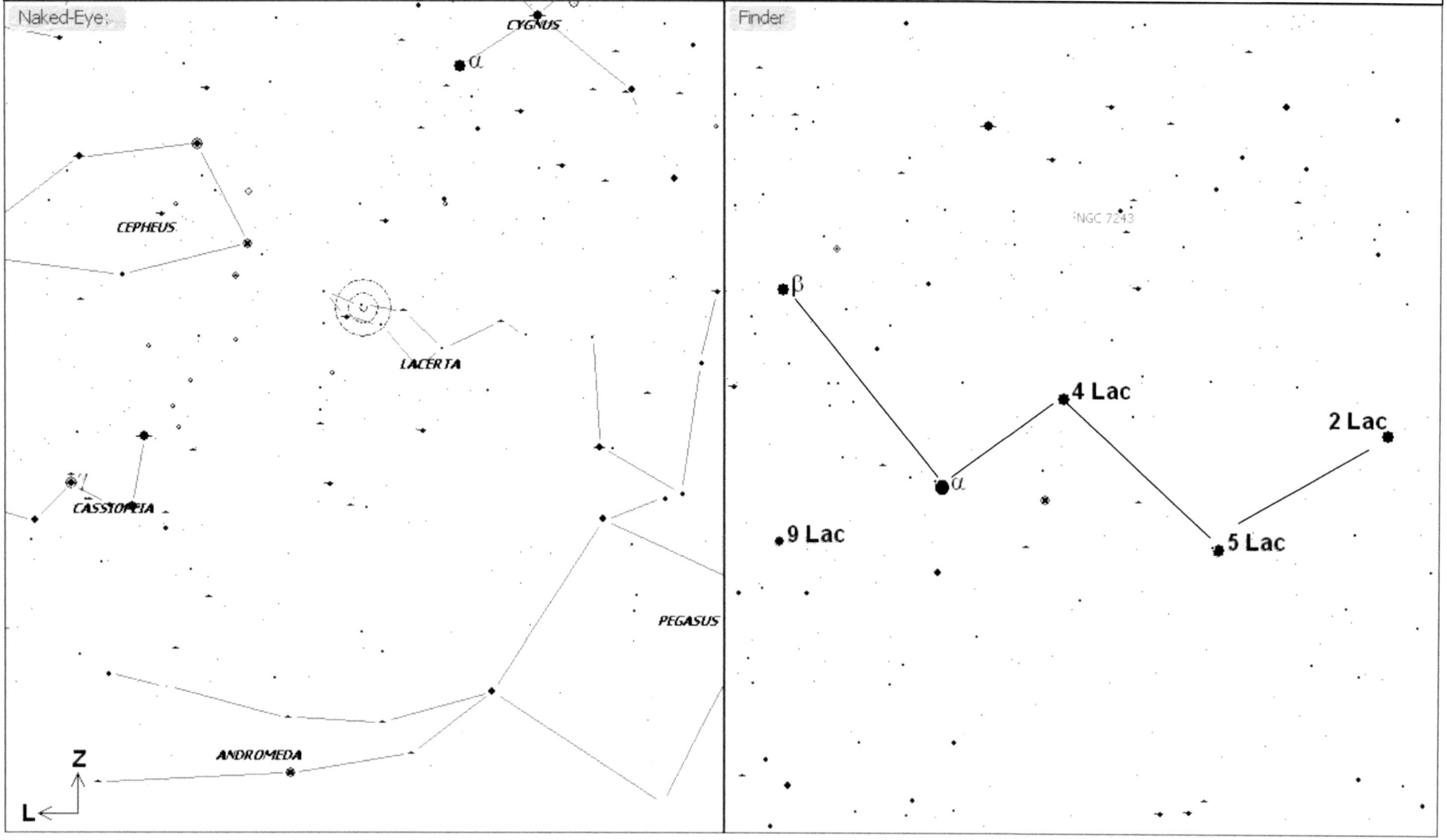

Naked-Eye:

Finder

7 x 50 Binoculars

Name	**The Tau Leonis Asterism**
Location	RA: 11h 27m 56s, Dec: +02° 51' 23" (2000) in Leo
Colorful Stars	83 Leo, Tau Leo, SAO 118883, 118885
Found With	8 x 50 Finderscope
Description	Golden Tau is the anchor of this five star binocular asterism in southeast Leo, near the ecliptic. The distance from 83 to SAO 118885 is 37'. Southeast Leo has several other asterisms.

Star chart showing: SAO 118894 A (82), 83 Leo A (65), B (77), C (99), τ A (49), B (81), SAO 118883 (93), SAO 118879 (72), SAO 118885 (86)

Principal Stars

83 Leo Multiple Star System
aka HR 4414, HD 99491, SAO 118864, Struve 1540, ADS 8162
Mag: H 6.62 to H 6.65, K0IV
Magnitude: 6.49, K0IV
Distance: 58 +/- 1.4 ly
AB: 6.49+7.7 mag, B=HD 99492
In 1991: PA 150° Sep 28.36"
AC: 6.49+9.9 mag,
In 1936: PA 188° Sep 90.30"
SAO 118879 Single Star
aka HD 99739, PPM 157890, HIP 55982
Magnitude: 7.24, F8
Distance: 170 +/- 8.6 ly

Tau Leo Multiple Star System
aka 84 Leo, HR 4418, HD 99648, SAO 118875, Struve 4019
Magnitude: 4.91, G8II
Distance: 620 +/- 99.0 ly
AB: 4.95+8.1 mag, B=HD 99649
In 1932: PA 176° Sep 91.10"
BC: 8.1+14.4 mag,
In 1910: PA 234° Sep 106.50"
AD: 4.95+10 mag, D=SAO 118883
In 1875: PA 92° Sep 764.80"
SAO 118883 Single Star
aka PPM 157899, BD +03 02507
Magnitude: 9.35, M0
SAO 118885 Single Star
aka HD 99798, PPM 157902, HIP 56029
Magnitude: 8.63, K0
Distance: 810 +/- 270.0 ly

Object data from SKYTOOLS™ 2 — used with permission.

The Tau Leonis Asterism / Leo / τ Leo / RA: 11h 27m 56s, Dec: +02° 51' 23"

Naked-Eye

Eyepiece: 37x, 80.0'

Finder

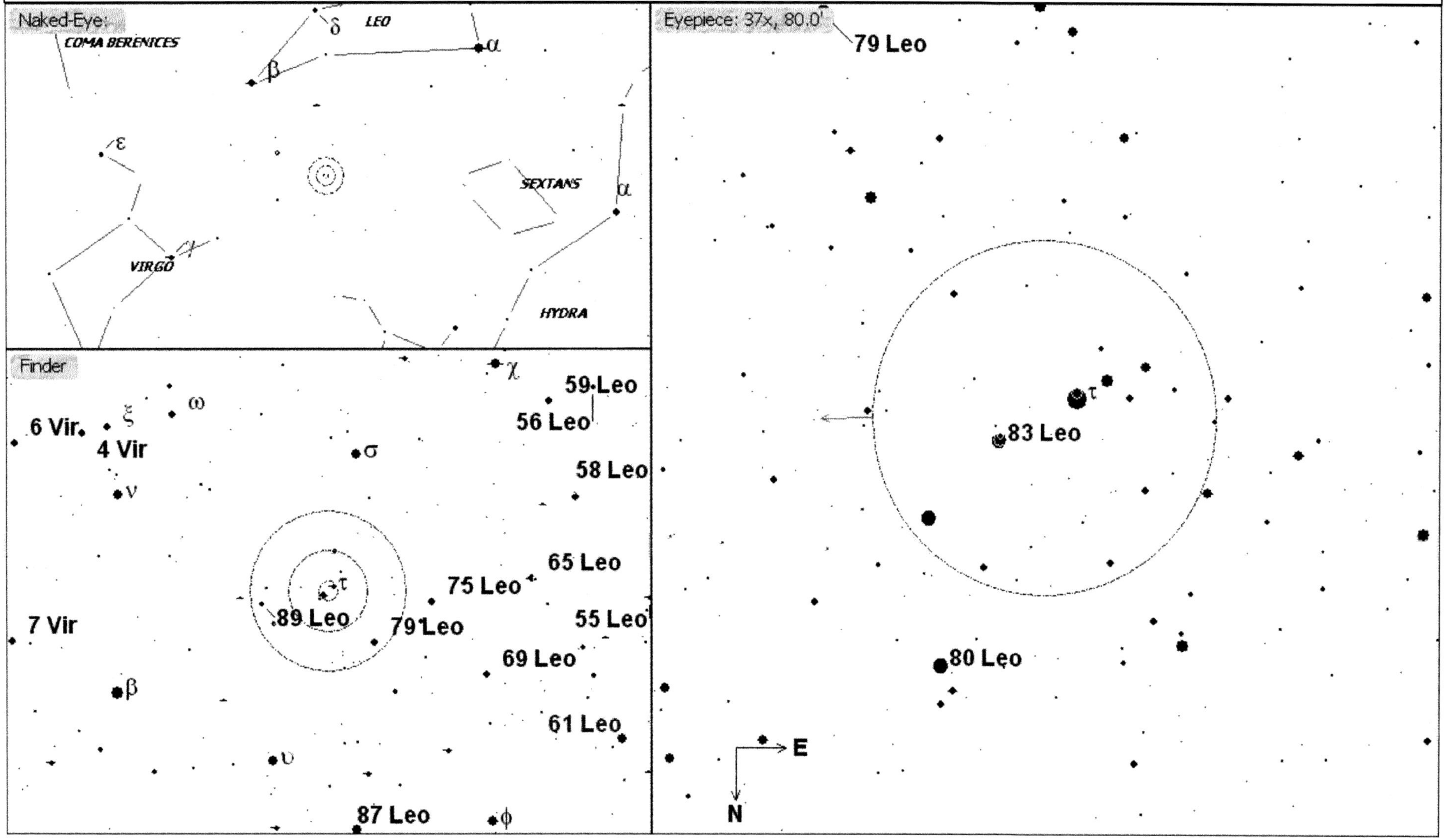

6" f/8 Reflector

Name	**The 22 Leo Minoris Asterism**
Location	RA 10h 15m 6s, Dec +31° 28' 05" in Leo Minor
Found With	8 x 50 Finderscope / April 5, 2003
Description	This asterism's shape resembles NGC 457 ("The Owl Cluster") in Cassiopeia, but is three times the size, and 22LMi is a single star in place of double Phi Cas, the two bight eyes of the Owl. The distance between 22 Lmi and SAO 61926 is 31'.

Principal Stars

SAO 61926 Single Star
aka HD 88436, PPM 75066, BD +32 01995
Magnitude: 8.47, F5

SAO 61934 Single Star
aka HD 88531, PPM 75078, HIP 50059
Magnitude: 8.38, F5
Distance: 360 +/- 42.0 ly

SAO 61936 Single Star
aka HD 88559, PPM 75080, BD +32 01997
Magnitude: 8.92, F8

SAO 61940 Single Star
aka PPM 75085, BD +32 01998, TYC 02510-0639-1
Magnitude: 9.55

SAO 61942 Single Star
aka PPM 75086, HIP 50136, BD +32 01999
Magnitude: 9.13, K0III
Distance: 960 +/- 360.0 ly

SAO 61943 Single Star
aka HD 88677, PPM 75088, HIP 50143
Magnitude: 7.62, F0
Distance: 310 +/- 26.0 ly

SAO 61944 Single Star
aka PPM 75090, HIP 50150, BD +31 02111
Magnitude: 9.10
Distance: 800 +/- 240.0 ly

SAO 61951 Single Star
aka PPM 75099, BD +31 02114
Magnitude: 8.81

SAO 61952 Single Star
aka PPM 75101, BD +32 02003
Magnitude: 8.87

*22 Lmi Single Star
aka HR 4014, HD 88786, SAO 61953
Magnitude: 6.47, G8III
Distance: 760 +/- 140.0 ly

Object data from SKYTOOLS™ 2 — used with permission.

The 22 Leo Minoris Asterism / Leo Minor / RA: 10h 15m 06s, Dec: +31° 28' 05"

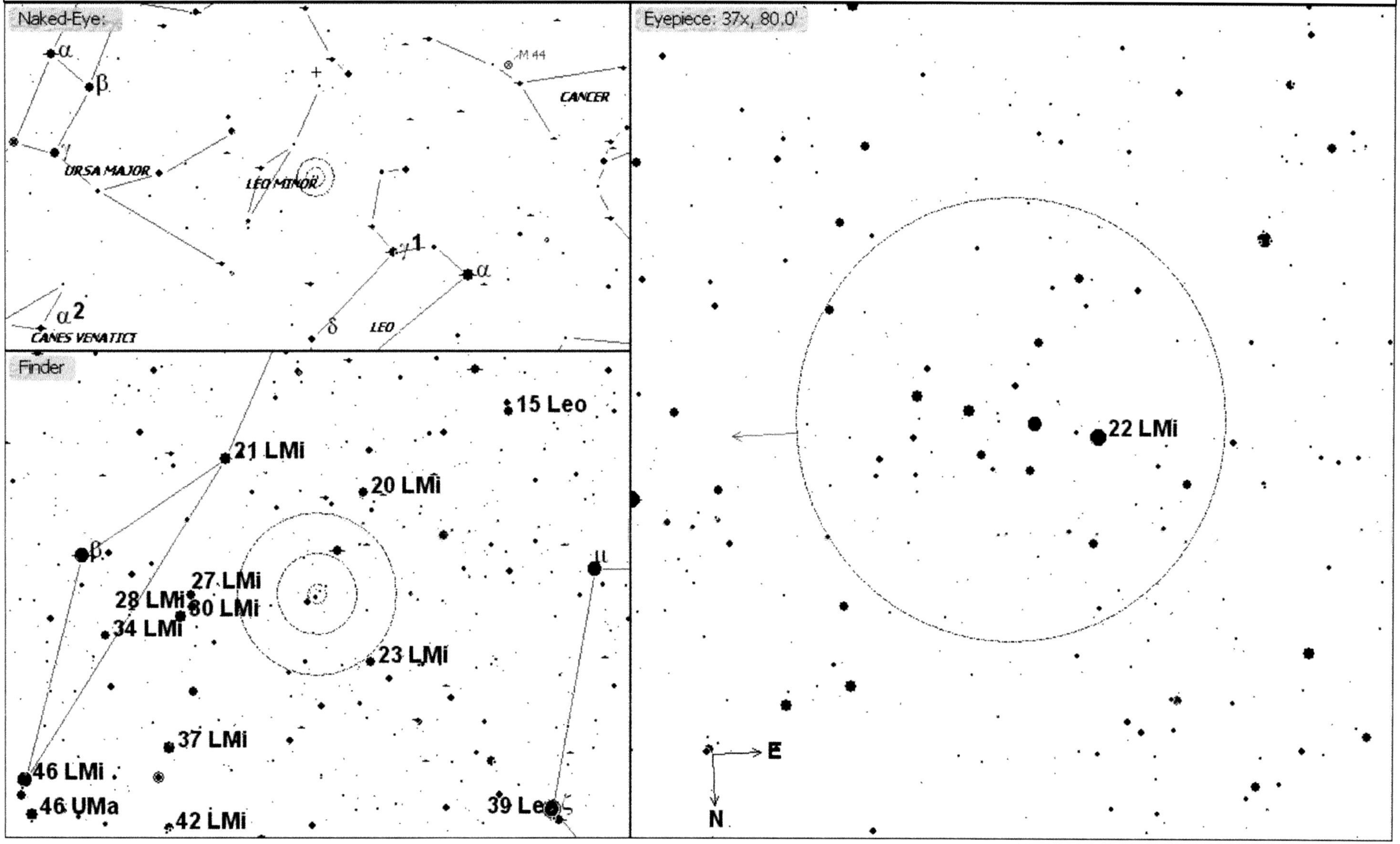

Eyepiece: 37x, 80.0'

Naked-Eye

Finder

6" f/8 Reflector

Name	**The 30 Leo Minoris Asterism**
Location	RA 10h 25m 55s, Dec +33° 47' 46" in Leo Minor
Colorful Stars	28 Lmi, SAO 62021 'A'
Found With	Direct Vision
Description	This triangle of 27, 28, and 30 LMi appears as a fuzzy spot to the naked eye. Closer inspection reveals the double SAO 62021. The brightest star within the quadrangle is SAO 62024. The distance between 27 and 30 Lmi is 35'.

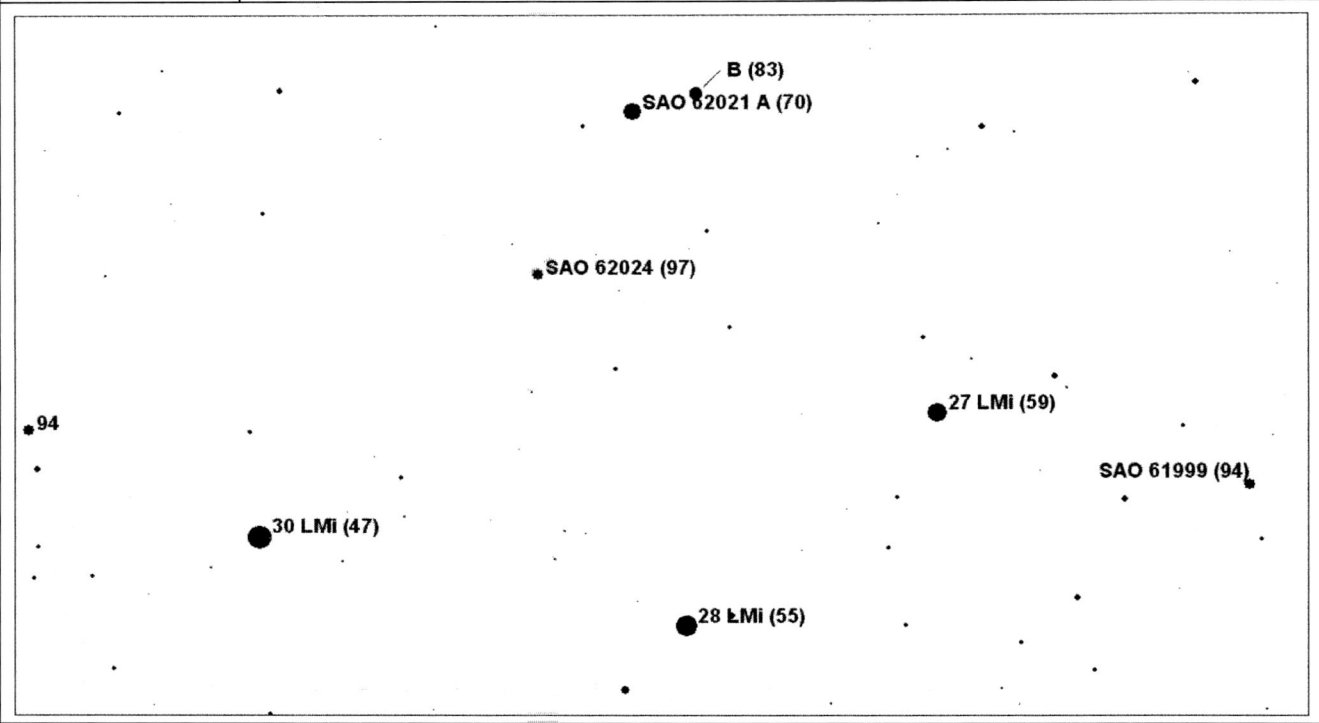

Principal Stars

27 LMi Single Star
aka HR 4075, HD 89904, SAO 62010
Magnitude: 5.89, A5
Distance: 230 +/- 14.0 ly

28 LMi Single Star
aka HR 4081, HD 90040, SAO 62019
Magnitude: 5.52, K0
Distance: 410 +/- 40.0 ly

SAO 62021 Multiple Star System
aka Otto Struve 4104, HD 90068, HIP 50951
Magnitude: 6.95, M0
Distance: 820 +/- 210.0 ly
AB: 6.95+8.3 mag, B=HD 90024
In 1875: PA 286° Sep 207"

SAO 62024 Single Star
aka PPM 75193, BD +34 02125,
TYC 02514-0844-1
Magnitude: 9.66

*30 LMi Single Star
aka HR 4090, HD 90277, SAO 62038
Magnitude: 4.74, F0
Distance: 210 +/- 11.0 ly

Object data from SKYTOOLS™ 2 — used with permission.

The 30 Leo Minoris Asterism / Leo Minor / RA: 10h 25m 55s, Dec:+33° 47' 46"

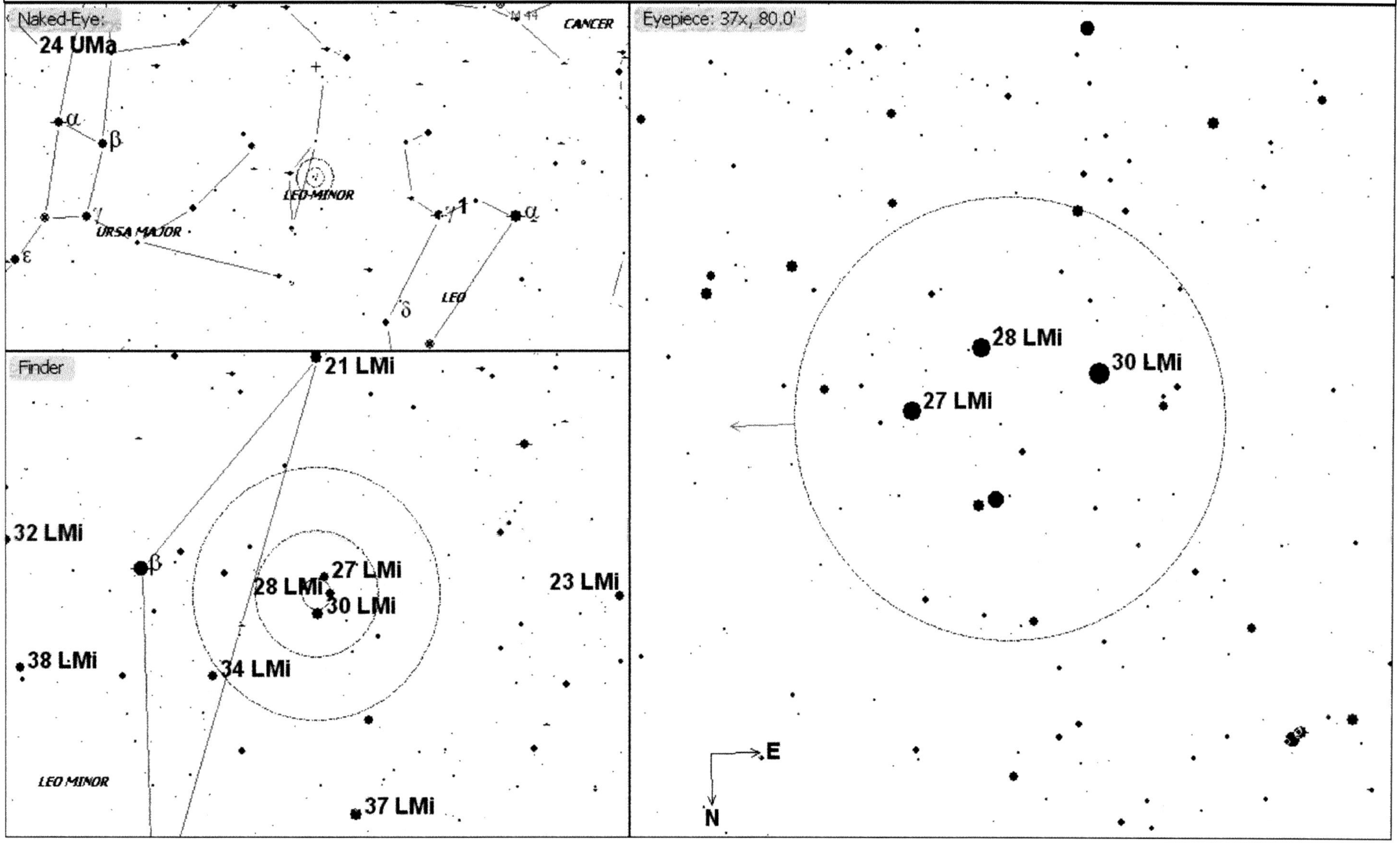

Eyepiece: 37x, 80.0'

Naked-Eye

Finder

6" f/8 Reflector

Name	**The 46 / 46 Asterism**
Location	RA: 10h 53m 19s, Dec +30° 12' 54" in Leo Minor and Ursa Major
Colorful Stars	46 LMi, SAO 62310, 46 UMa, SAO 62318
Found With	Direct Vision
Description	This asterism is seen with the naked eye as a fuzzy object between the rear feet of Ursa Major. Telescopic inspection reveals a meandering chain of stars crossing the border between Leo Minor and Ursa Major. Two other close stars with the same Flamsteed number are 1 Sge and 1 Vul.

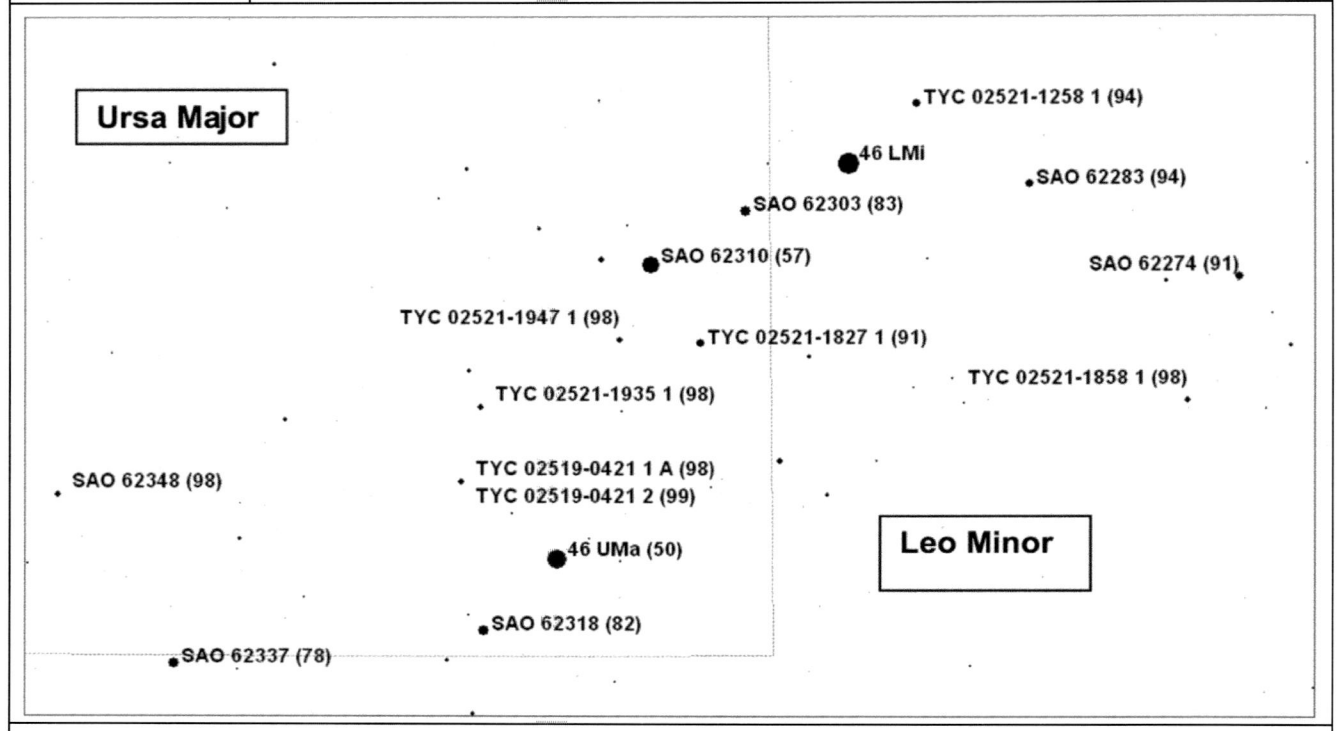

Principal Stars

TYC 02521-1258-1 Single Star in Leo Minor
aka BD +35 02192
Magnitude: 9.37
*46 LMi Variable Star
aka, HR 4247, HD 94264, SAO 62297
Variability: Mag: V 3.79 to V 3.84, K0
Distance: 98 +/- 2.3 ly
SAO 62303 Single Star in Ursa Major
aka HD 94383, PPM 75542, HIP 53305
Magnitude: 8.35, G0
Distance: 230 +/- 18.0 ly
SAO 62310 Single Star in Ursa Major
aka HR 4256, HD 94497
Magnitude: 5.73, K0
Distance: 300 +/- 22.0 ly
SAO 62313 Single Star in Ursa Major
aka HD 94550, PPM 75555, HIP 53404
Magnitude: 10.08, G0
Distance: 1100 +/- 570.0 ly

TYC 02521-1827-1 Single Star in Ursa Major
aka SI 232526
Magnitude: 9.14
TYC 02521-1947-1 Single Star in Ursa Major
aka SI 232535
Magnitude: 9.79
TYC 02521-1935-1 Single Star in Ursa Major
aka SI 232533
Magnitude: 9.84
TYC 02519-0421-1 Multiple Star in Ursa Major
aka HD 94728, HIP 53488
Magnitude: 9.82 Distance: 630 +/- 220.0 ly
AB: 9.82+12.78 mag
In 1991: PA 94° Sep 1.76"
46 Uma Single Star
aka HR 4258, HD 94600, SAO 62314
Magnitude: 5.02, K0
Distance: 240 +/- 14.0 ly
SAO 62318 Single Star in Ursa Major
aka HD 94700 , PPM 75563
Magnitude: 8.19, K2

Object data from SKYTOOLS™ 2 — used with permission.

The 46 / 46 Asterism / Leo Minor / 46 LMi / RA: 10h 53m 19s, Dec: +34° 12' 54"

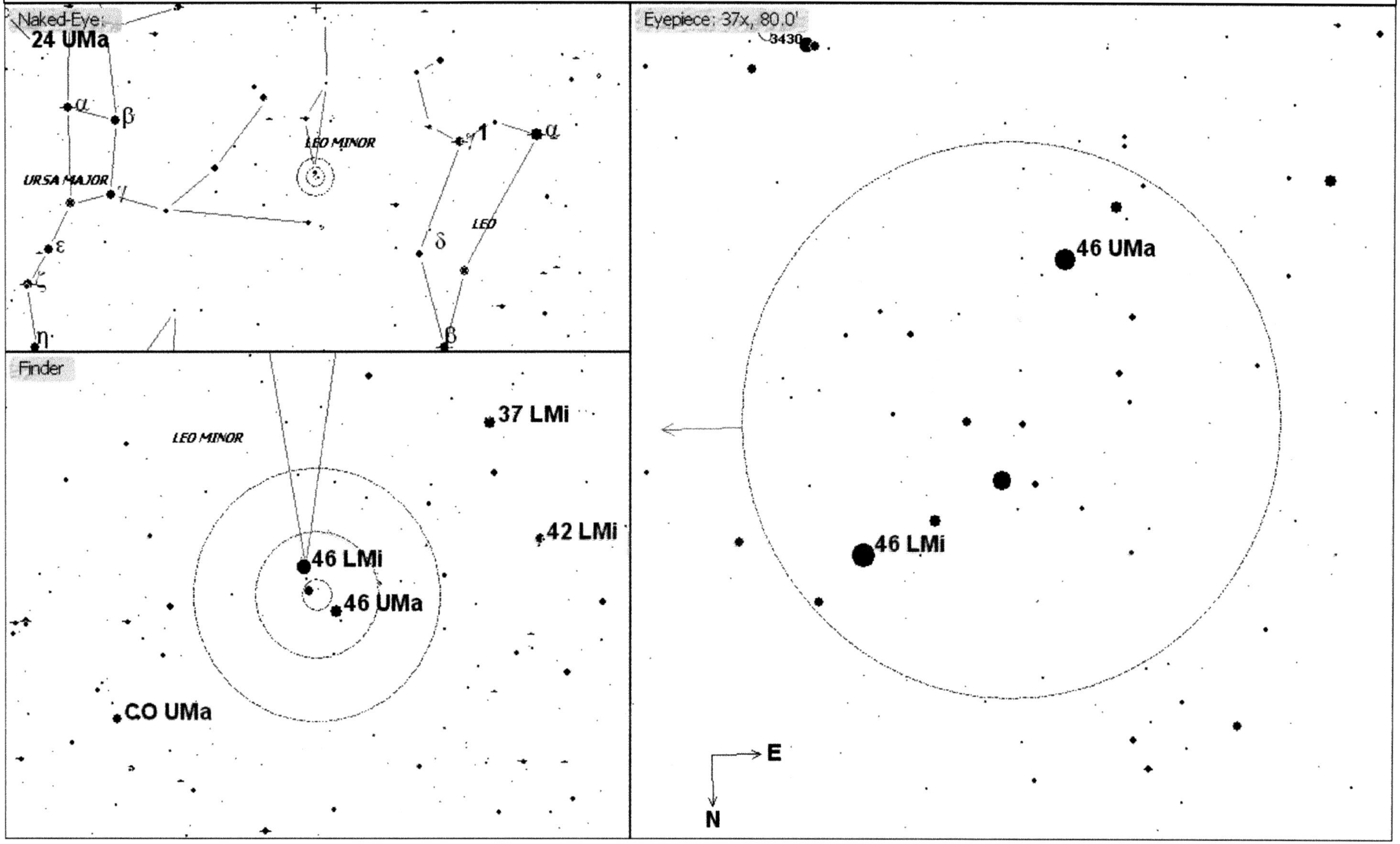

Eyepiece: 37x, 80.0'

Naked-Eye
24 UMa

Finder

6" f/8 Reflector

Name	**The Libra - Hydra Asterism**
Location	RA 14h 46m 00s, Dec –25° 26' 35" In Hydra and Libra
Found With	Direct Vision
Description	This asterism appears as a fuzzy spot to the naked eye. It lies right on the border of Libra and Hydra, in the narrow easternmost end of the monster. The slightest magnification reveals five stars in a symmetrical formation. The boomerang shape of this asterism reminds me of the Stealth Bomber plane. The distance between 4 Librae and 57 Hydrae is about two deg.

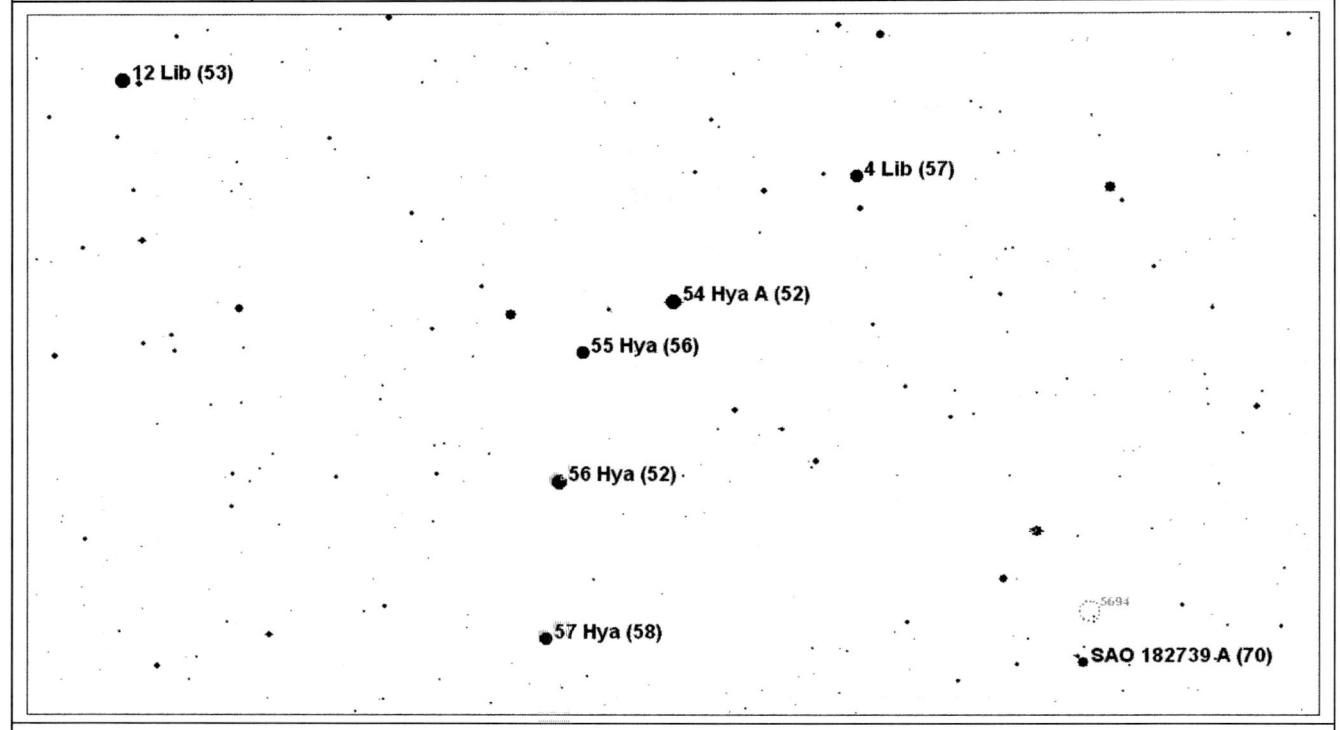

Principal Stars

4 Lib Single Star
aka HR 5484, HD 129433, SAO 182795
Magnitude: 5.70, B9
Distance: 410 +/- 40.0 ly

*54 Hya Multiple Star System
aka HR 5497, HD 129926, SAO 182856,
H 97, ADS 9375
Magnitude: 5.15, F0
Distance: 99 +/- 2.8 ly
AB: 5.15+7.25 mag, B=HD 129926;
In 1991: PA 124° Sep 8.32"

55 Hya Single Star
aka HR 5514, HD 130158, SAO 182875
Magnitude: 5.61, B9IV
Distance: 610 +/- 86.0 ly

56 Hya Single Star
aka HR 5516, HD 130259, SAO 182882
Magnitude: 5.23, G8III
Distance: 330 +/- 26.0 ly

57 Hya Single Star
aka HR 5517, HD 130274, SAO 182883
Magnitude: 5.76, B9
Distance: 430 +/- 49.0 ly

Object data from SKYTOOLS™ 2 — used with permission.

The Libra / Hydra Asterism / 54 Hya / RA: 14h 46m 00s, Dec: -25° 26' 35"

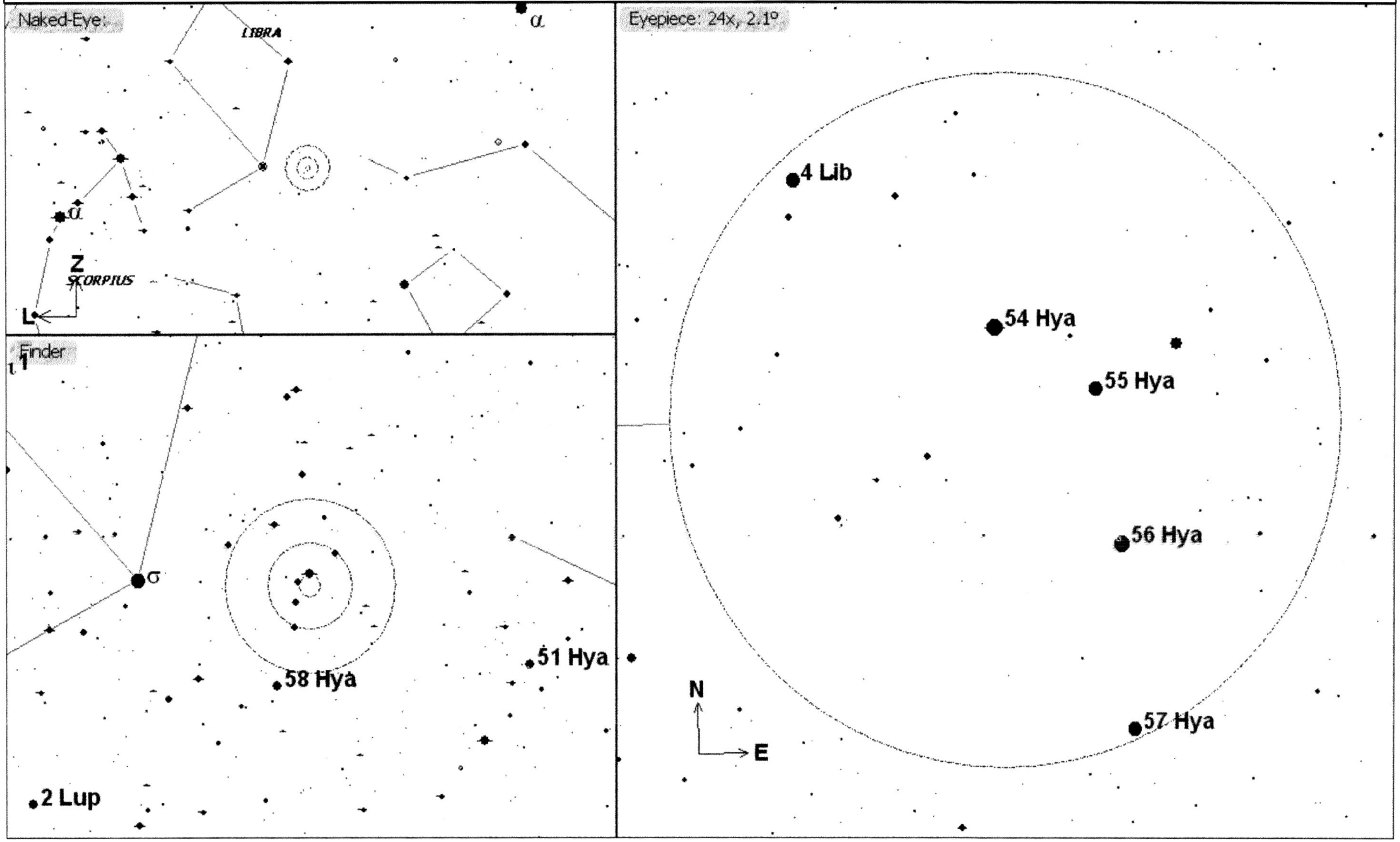

Eyepiece: 24x, 2.1°

80mm f/6 Refractor

123

Name	**The Libra Pseudo-Cluster**
Location	RA 14h 23m 43s, Dec −12° 56' 33" in Libra
Colorful Stars	SAO 158529
Found With	8 x 50 Finderscope
Description	Impression when seen with 8x50 finder: "Wow! What's that? This asterism reminds me of M29 in Cygnus. This asterism is about 5' in diameter.

Principal Stars

SAO 158529   Single Star
aka HD 126055, PPM 228715, BD -12 04032, TYC 05570-0750 1, SI 584571
Magnitude: 9.91, K1III
SAO 158531   Single Star
aka HD 126068, PPM 228720, BD -12 04034, TYC 05570-0571 1, SI 584560
Magnitude: 9.59, G0
SAO 158532   Multiple Star System
aka HD 126083, PPM 228721, HIP 70357
Magnitude: 7.82, A1IV
Distance: 620 +/- 110.0 ly
AB: 7.82+10.17 mag
In 1991: PA 245° Sep 0.20"

TYC 05570-0815-1   Single Star
aka PPM 717546, SI 584578
Magnitude: 9.84
*SAO 158534   Single Star
aka HD 126101, PPM 228724, HIP 70365, BD -12 04037, TYC 05570-0833
Magnitude: 7.66, F5
Distance: 360 +/- 40.0 ly
TYC 05570-0408-1 Single Star
aka SI 584550
Magnitude: 10.88

Object data from SKYTOOLS™ 2 — used with permission.

The Libra Pseudo-Cluster / Libra / SAO 158534 / RA: 14h 23m 43s, Dec: -12° 58' 49"

Naked-Eye

Finder

Eyepiece: 37x, 80.0'

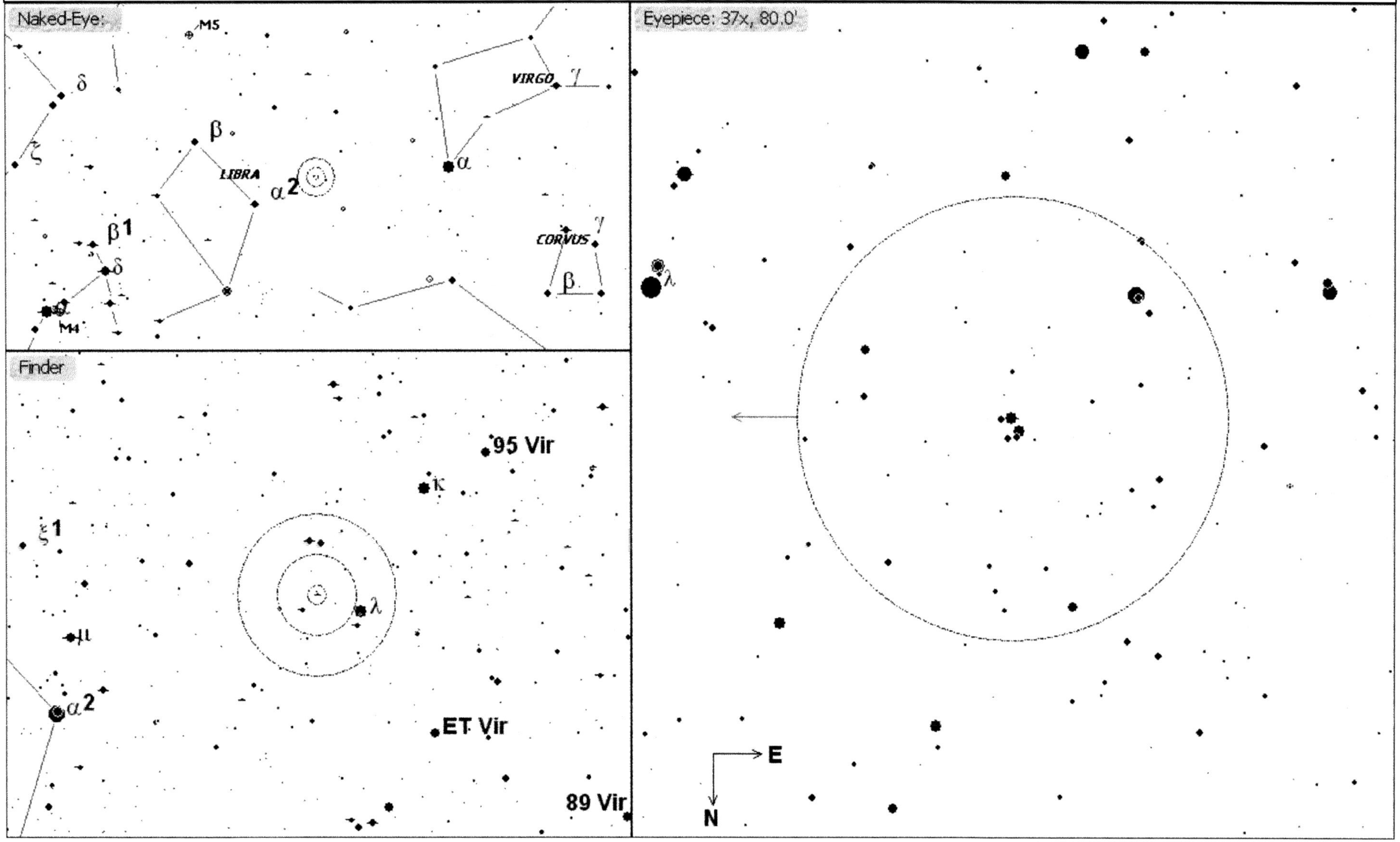

Eyepiece View: λ Vir at left

6" f/8 Reflector

125

Name	**Theta (θ) Lupi Asterism**
Location	RA: 16h 06m 36s, Dec –36° 48' 08" in Lupus
Found With	8 x 50 Finder
Description	Three bright stars, two white, one orange. The distance from Theta to SAO 207348 is 13'.

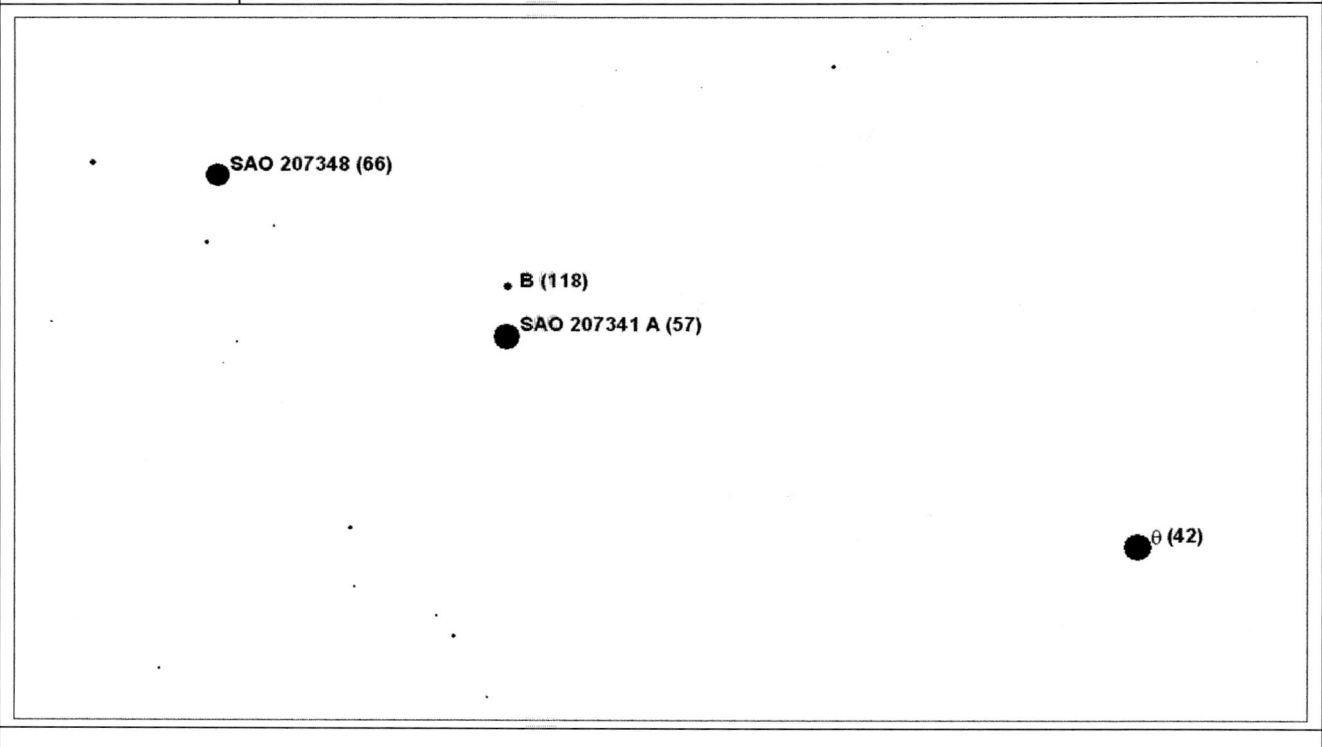

Principal Stars

***Theta Lupi** Single Star
aka HR 5987, HD 144294, SAO 207332, GC 21625, TYC 07342-1315 1
Magnitude: 4.23, B2
Distance: 410 +/- 43.0 ly
SAO 207341 Multiple Star System
aka HR 5991 HD 144415, PPM 294675, HIP 78970, HJ 4831
Magnitude: 5.72, F0
Distance: 170 +/- 18.0 ly
AB: 5.72+11.8 mag
In 1900: PA 358° Sep 40.70"
SAO 207348 Single Star
aka HD 144475, PPM 294682 SAO 207348, HIP 79000
Magnitude: 6.64, K1III
Distance: 470 +/- 54.0 ly

Object data from SKYTOOLS™ 2 — used with permission.

The Theta (θ) Lupi Asterism / Lupus / RA: 16h 06m 36s, Dec: -36° 48' 08"

Naked-Eye

Finder

Eyepiece: 37x, 80.0'

E / N

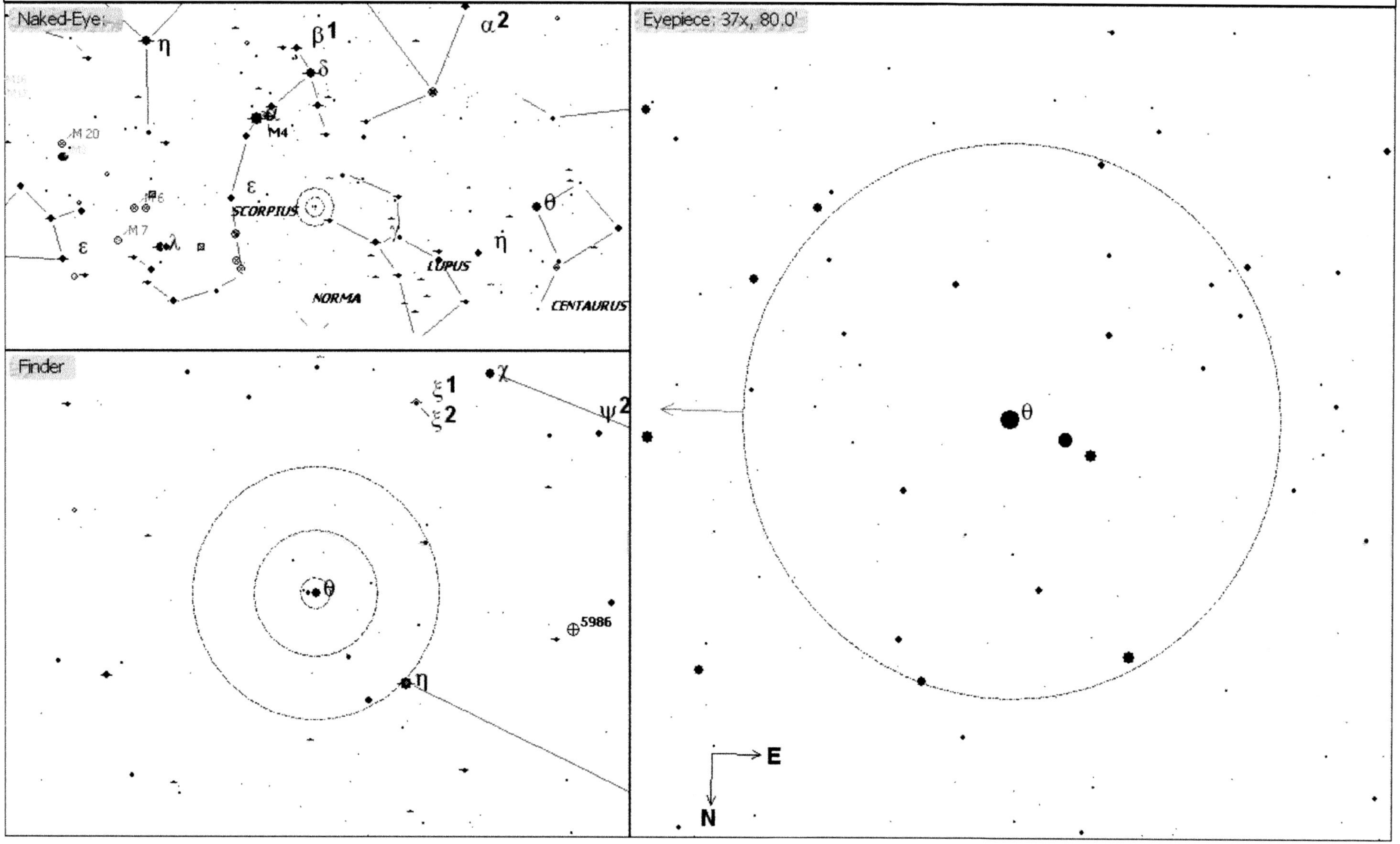

6" f/8 Reflector

Name	**Yed Prior Asterism**
Location	RA 16h 14m 21s, Dec: -3° 41' 40" in Ophiuchus
Colorful Stars	Yed Prior, SAO 141043
Found With	6" Reflector
Description	A bright asterism about 3/4 degree SE of Yed Prior. The orange SAO 141043 complements the color of Yed Prior.

Principal Stars

Yed Prior Multiple Star System
aka Delta Oph, 1 Oph, HR 6056,
HD 146051, SAO 141052
AB: 2.73+13.4 mag
In 1908: PA 294° Sep 65.50"
Magnitude: 2.75, M0.5 III
Distance: 170 +/- 9.1 ly
SAO 141048 Single Star
aka HD 146013, PPM 199438, HIP 79567
Magnitude: 7.40, A0
Distance: 580 +/- 120.0 ly

SAO 141043 Single Star
aka HD 145960, PPM 199433
Magnitude: 7.93, K0
SAO 141036 Single Star
aka HD 145876, PPM 199422
Magnitude: 6.96, A5
Distance: 270 +/- 25.0 ly
SAO 141031 Single Star
aka HR 6041, HD 145788
Magnitude: 6.26, A1
Distance: 560 +/- 100.0 ly

Object data from SKYTOOLS™ 2 — used with permission.

The Yed Prior Asterism / Ophiuchus / Delta (δ) Oph / RA: 16h 14m 21s, Dec: -03° 41' 40"

Eyepiece: 60x, 50.0'

Naked-Eye

Finder

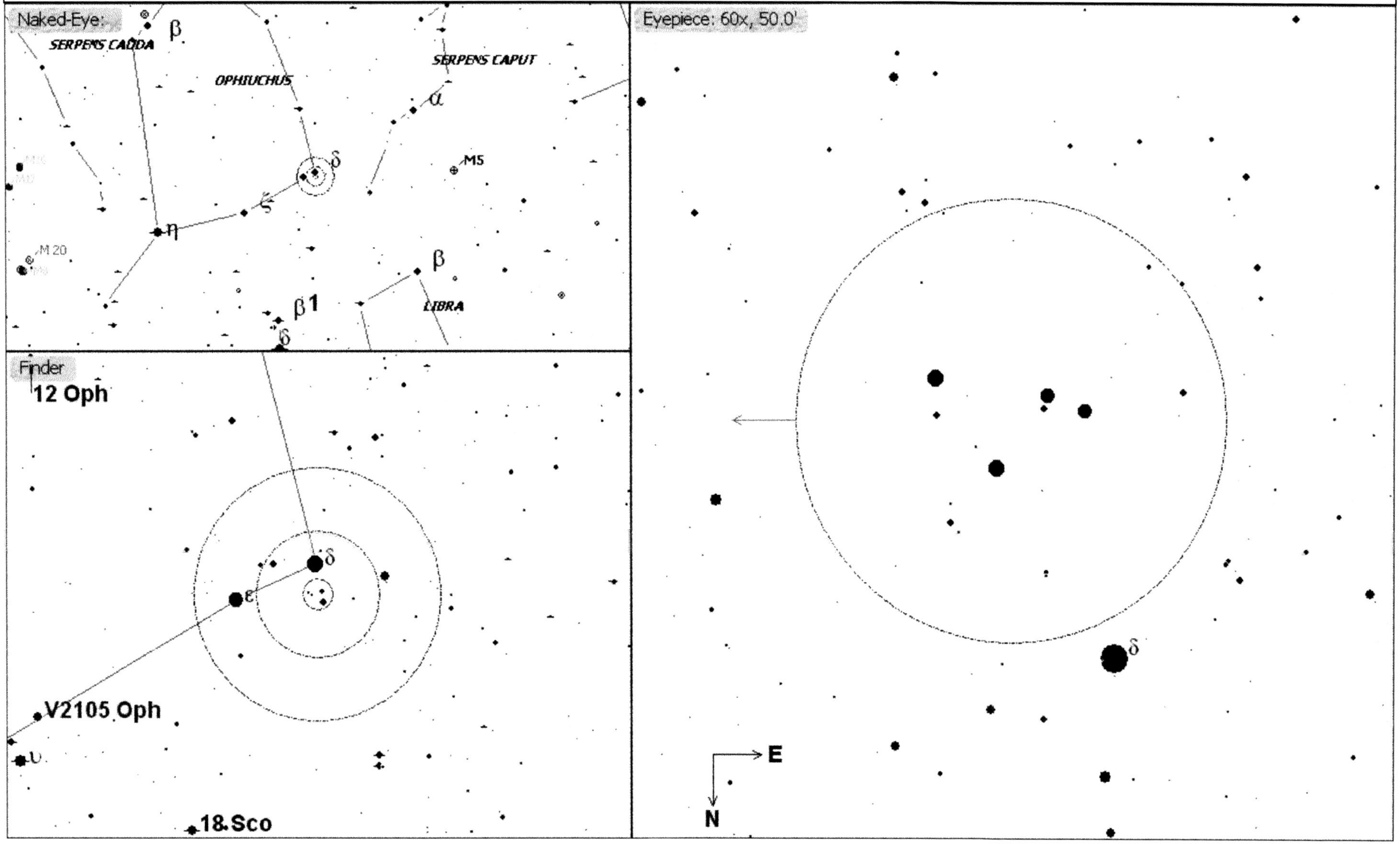

6" f/8 Reflector

Name	**The Kappa Ophiuchi Asterism**
Location	RA: 16h 59m 30s, Dec: +09° 42' 12"
Colorful Stars	SAO 121991
Found With	11 x 70 Binoculars, 7-14-04
Description	This small asterism is located 1 degree north-east of Kappa Ophiuchi. About 15' separates SAO 121980 and 121991.

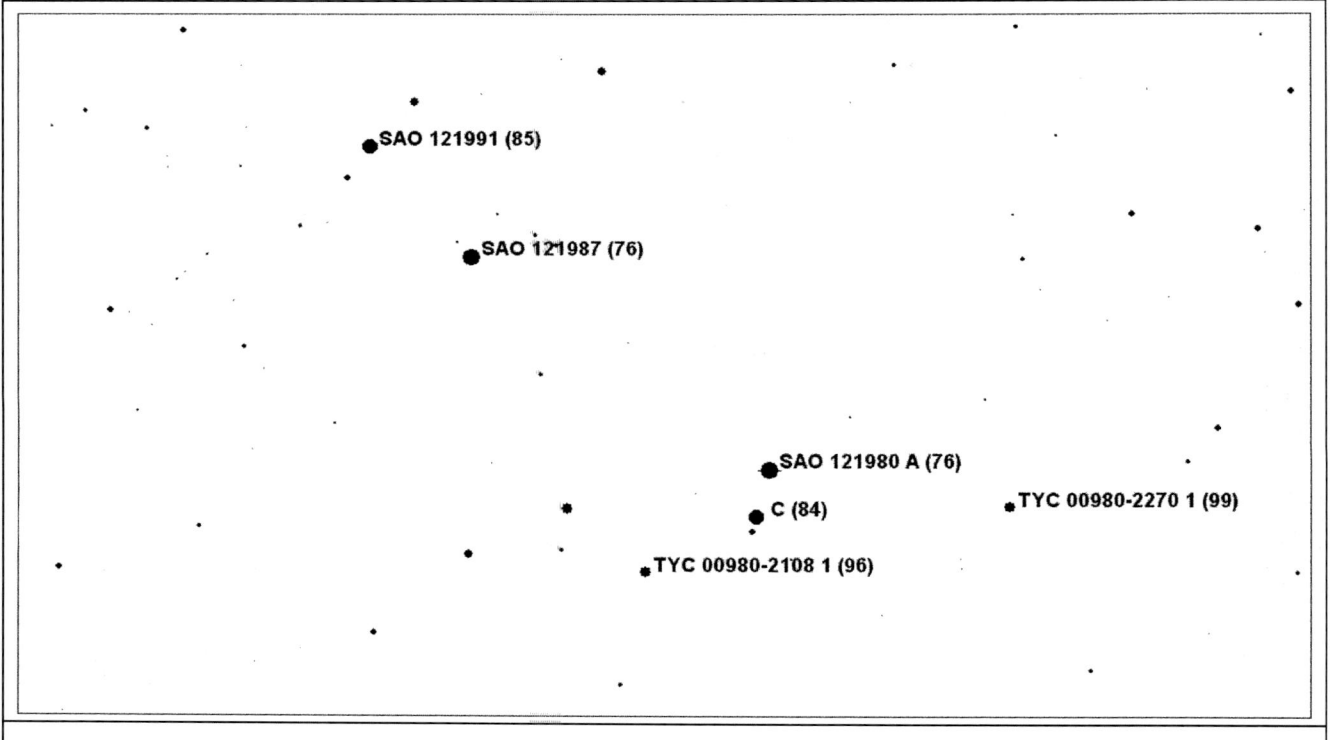

Principal Stars

TYC 00980-2270-1 Single Star
aka PPM 163344, BD +09 03301
Magnitude: 9.88
*SAO 121980 Multiple Star System
aka HD 153475, PPM 163353, HIP 83143, ADS 10295, BU 1298
Magnitude: 7.96, F0
Distance: 630 +/- 290.0 ly
AB: 7.96+8.83 mag
In 1991: PA 124° Sep 0.43"
AC: 7.96+8.4 mag, STT4150, C=HD 153476
In 1874: PA 165° Sep 77.00"
CD: 8.4+12.4 mag, Struve 2111
In 1901: PA 164ª Sep 24.00"

TYC 00980-2108-1 Single Star
aka PPM 163358, BD +09 03307,
Magnitude: 9.55
SAO 121987 Single Star
aka HD 153562, PPM 163361, HIP 83178
Magnitude: 7.59, A0
Distance: 770 +/- 150.0 ly
SAO 121991 Single Star
aka HD 153603, PPM 163363
Magnitude: 8.52, K0

Object data from SKYTOOLS™ 2 — used with permission.

The Kappa (κ) Ophiuchi Asterism / Ophiuchus / RA: 16h 59m 30s, Dec: +09° 42' 12"

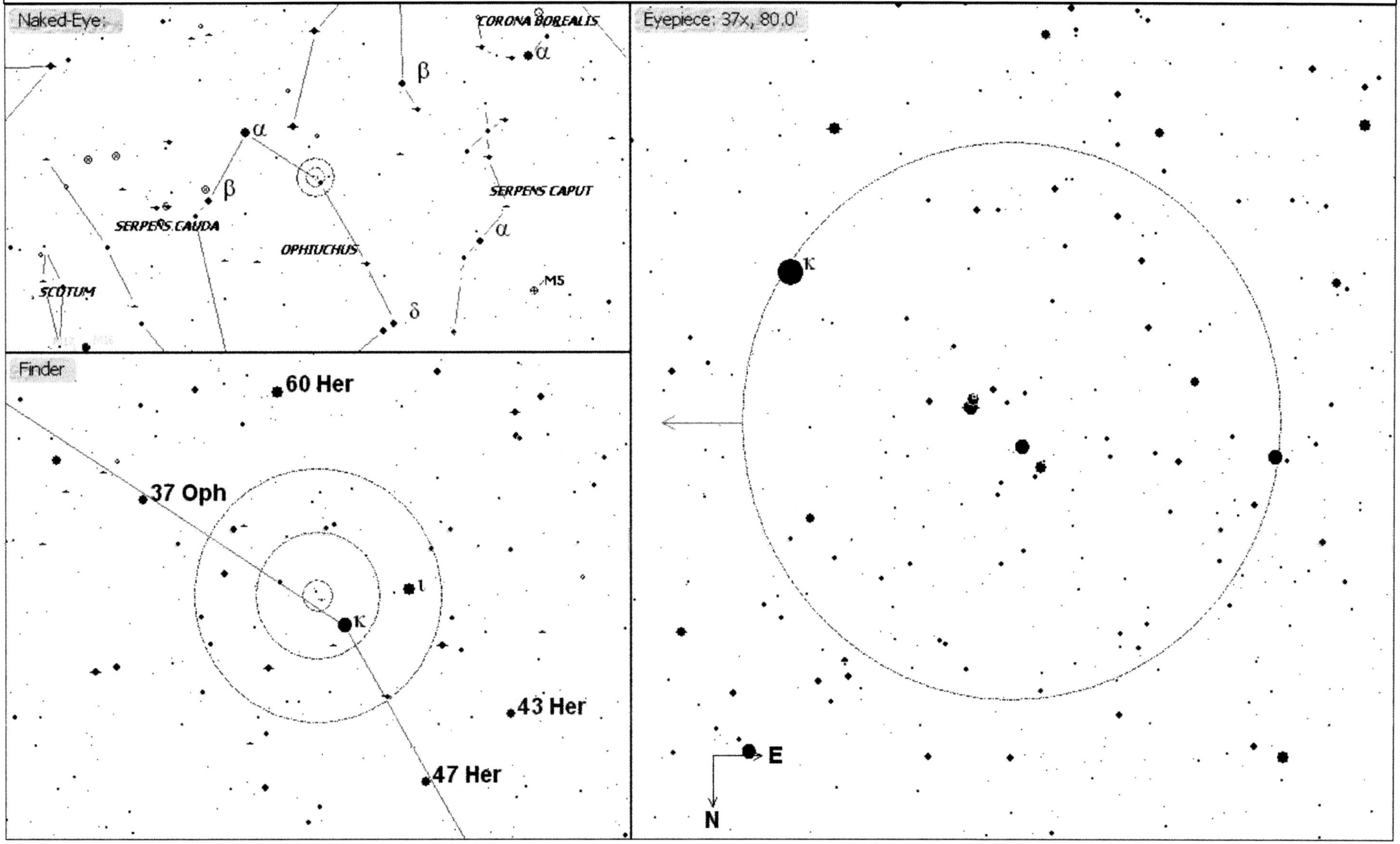

Eyepiece: 37x, 80.0

Naked-Eye

Finder

- 60 Her
- 43 Her
- 47 Her
- 37 Oph
- κ
- ι

6" f/8 Reflector

Name	**Enif Asterism**
Location	RA: 21h 42m 52s, Dec: +10° 30' 43" (2000) in Pegasus
Found With	7" f/15 Maksutov
Description	Located about 3/4 of a degree NW of Enif (ε Peg). I found this asterism while star hopping from Enif to M15. It's shape reminds me of a child's drawing of a rocket, or the Teapot of Sagittarius. It's faint but distinct.

Principal Stars

GSC 01125-1504 Single Star
Magnitude: 11.99

TYC 01125-1000-1 Single Star
aka SI 1270978
Magnitude: 11.23

TYC 01125-1940-1 Single Star
aka SI 100458
Magnitude: 10.72

TYC 01125-0784-1 Single Star
aka SI 1270966
Magnitude: 11.62

***SAO 107343** Single Star
aka HD 206589 , PPM 140529
Magnitude: 9.29, A0

SAO 107344 Single Star
aka HD 206590 , PPM 140530
Magnitude: 9.54, A0

SAO 107345 Single Star
aka PPM 140535, BD +09 04884
Magnitude: 9.68, F8

SAO 107347 Single Star
aka HD 206619, PPM 140538,
Magnitude: 9.41, A2

TYC 01125-1174-1 Single Star
aka SI 1270983
Magnitude: 11.78

SAO 107351 Single Star
aka PPM 140546, BD +09 04888
Magnitude: 9.72, F8

TYC 01125-0696-1 Single Star
aka PPM 140550, BD +09 04889
Magnitude: 10.36

Object data from SKYTOOLS™ 2 — used with permission.

Enif Asterism / Pegasus / ε Pegasi / RA: 21h 42m 52s, Dec: +10° 30' 43"

Eyepiece: 150x, 20.0'

Naked-Eye

Finder

20 Peg 17 Peg
4 Peg
7 Peg
M15

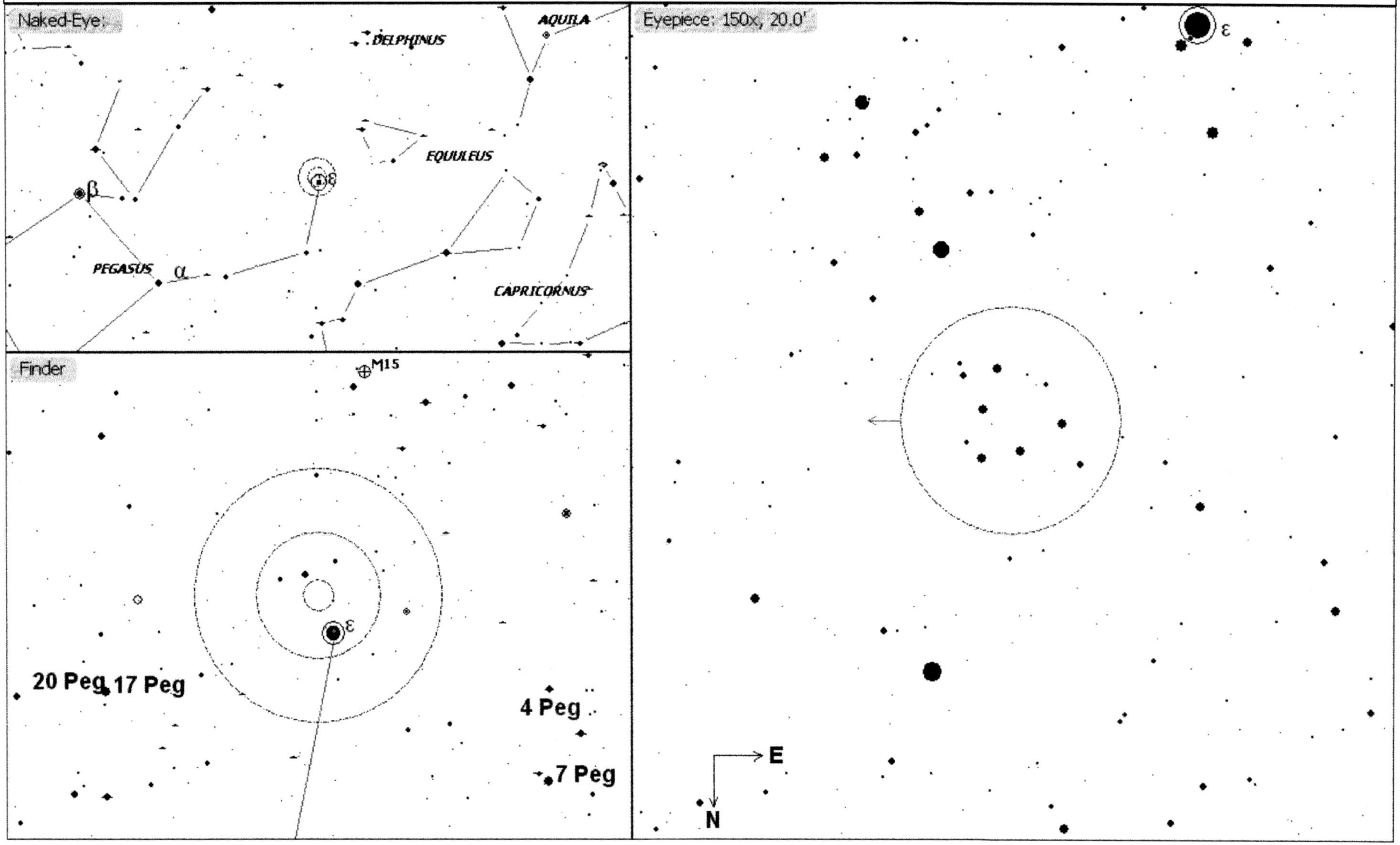

6" f/8 Reflector

Name	**The 35 Pegasi Asterism**
Location	RA: 22h 27m 5s, Dec: +04° 41' 44" (2000) in Pegasus
Colorful Stars	35 Peg
Found With	Direct Vision
Description	The trio of 34, 35, and 37 Peg appear as a distinctly hazy spot in the sky just north of the "Water Jar" of Aquarius. All three are multiple stars. The distance from 34 to 37 Peg is 50 arc-minutes.

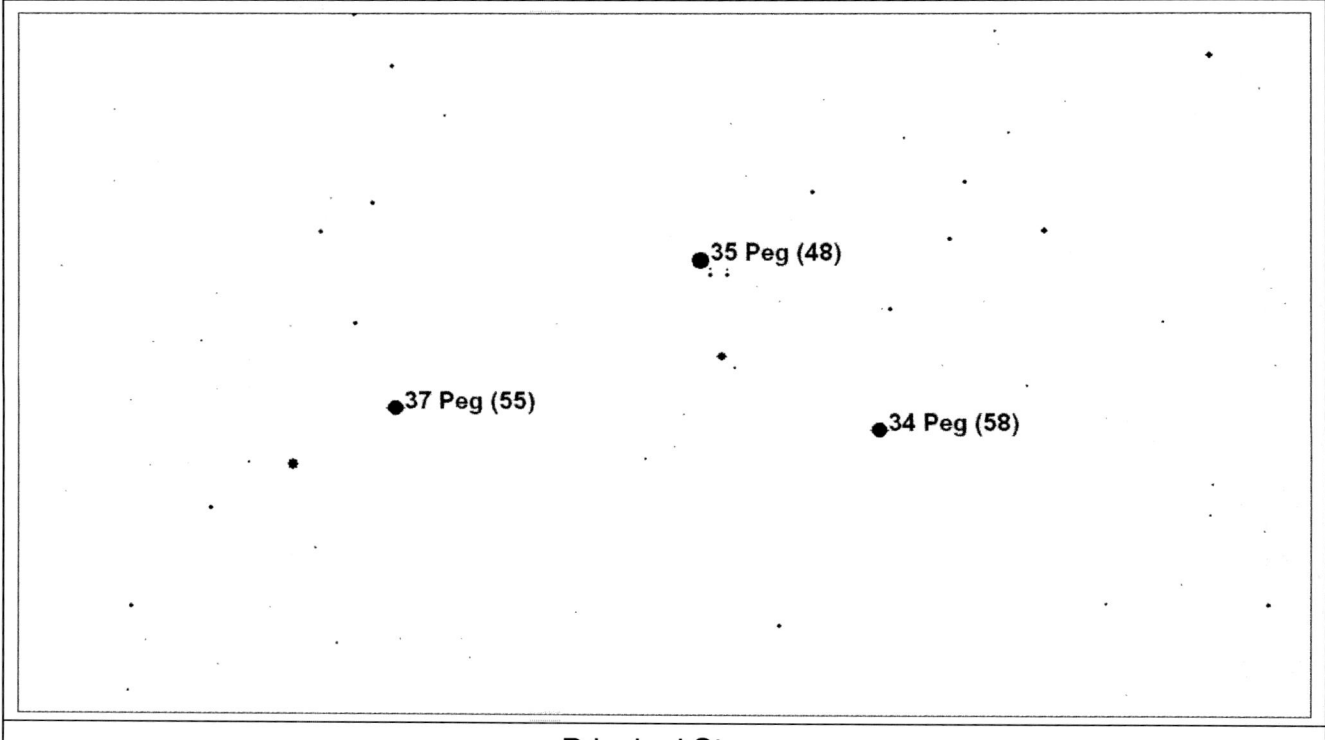

Principal Stars

34 Peg   Multiple Star System
aka HR 8548, HD 212754, SAO 127529,
BU 290, ADS 15935
Magnitude: 5.76, F7
Distance: 130 +/- 8.4 ly
AB: 5.76+12.3 mag
In 1937: PA 224° Sep 3.5"
AC: 5.76+12.8 mag
In 1924: PA 272° Sep 103.3"

37 Peg   Multiple Star System
aka HR 8566, HD 213235, SAO 127551,
Struve 2912, ADS 15988
Magnitude: 5.53, A7
Distance: 170 +/- 11.0 ly
AB: 5.51+7.37 mag
Reliable Orbit: P=140.0 yr, a=0.75"
PA 118° Sep 0.63" (2005.4)

35 Peg   Multiple Star System
aka HR 8551, HD 212943, SAO 12754
Magnitude: 4.80, K0
Distance: 160 +/- 6.1 ly
AB: 4.78+9.9 mag
In 1908: PA 210° Sep 98.30"
AC: 4.78+9.8 mag
In 1892: PA 241° Sep 181.50"
Close up view of 35 Peg:

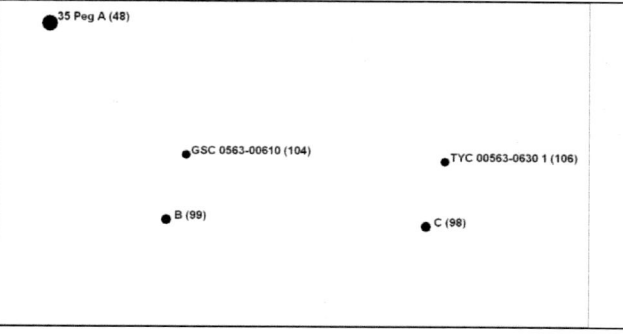

All information from SKYTOOLS™ 2 — used with permission.

The 35 Pegasi Asterism / Pegasus / RA: 22h 27m 51s, Dec: +04° 41' 44"

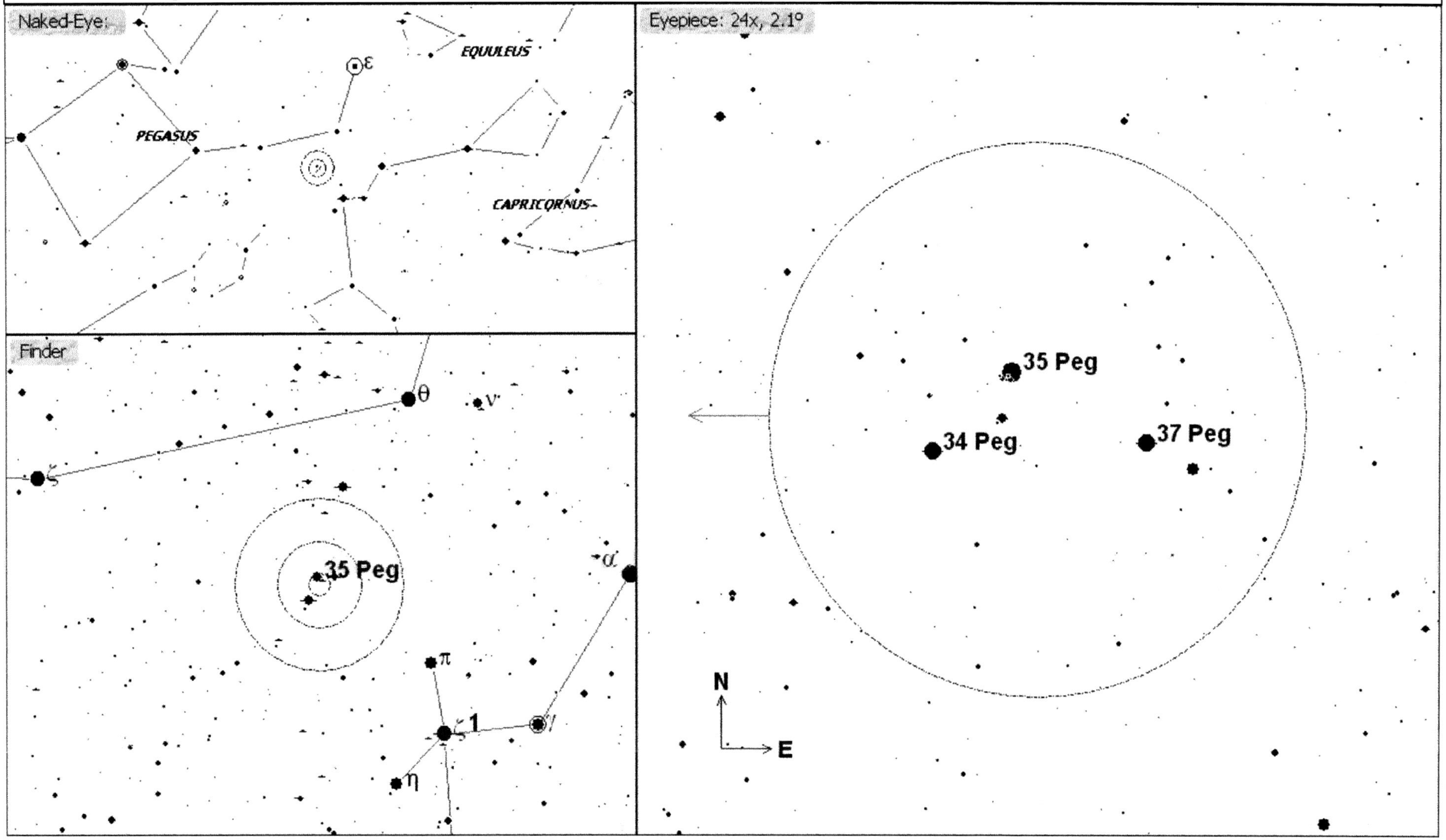

Eyepiece: 24x, 2.1°

35 Peg
37 Peg
34 Peg

N E

Naked-Eye
PEGASUS
EQUULEUS
CAPRICORNUS

Finder
35 Peg
θ

80mm f/6 Refractor

Name	**The Kappa Pegasi Asterism**
Location	RA: 21h 44m 39s, Dec: +25° 38' 42" (2000) in Pegasus
Colorful Stars	SAO 89976
Found With	8 x 50 Finder
Description	This asterism is easily seen in a small glass. This asterism is only 3 degrees north of the 12 Pegasi asterism. Both are distinctive star groups in the finderscope. The distance from Kappa to SAO 89976 is 24'.

Principal Stars

***Kappa Peg**   Multiple Star System
aka 10 Peg, HR 8315, HD 206901,
BU 989, ADS 15281
Magnitude: 4.12, F4IV
Distance: 120 +/- 3.6 ly
AB: 4.14+5.04 mag
PA 266° Sep 0.16" (2004.7)
Definitive Orbit: P=11.6 yr, a=0.25"
AC: 4.14+10.8 mag, AC = Struve 2824
In 1958: PA 292° Sep 13.80"

SAO 89971   Single Star
aka HD 207071, PPM 113547, HIP 107462
Magnitude: 6.57, B8
Distance: 650 +/- 100.0 ly
SAO 89976 Single Star
aka HD 207134, HR 8325, PPM 113556
Magnitude: 6.28, K3
Distance: 370 +/- 33.0 ly

Object data from SKYTOOLS™ 2 — used with permission.

The Kappa (κ) Pegasi Asterism / Pegasus / RA: 21h 44m 39s, Dec: +25° 38' 42"

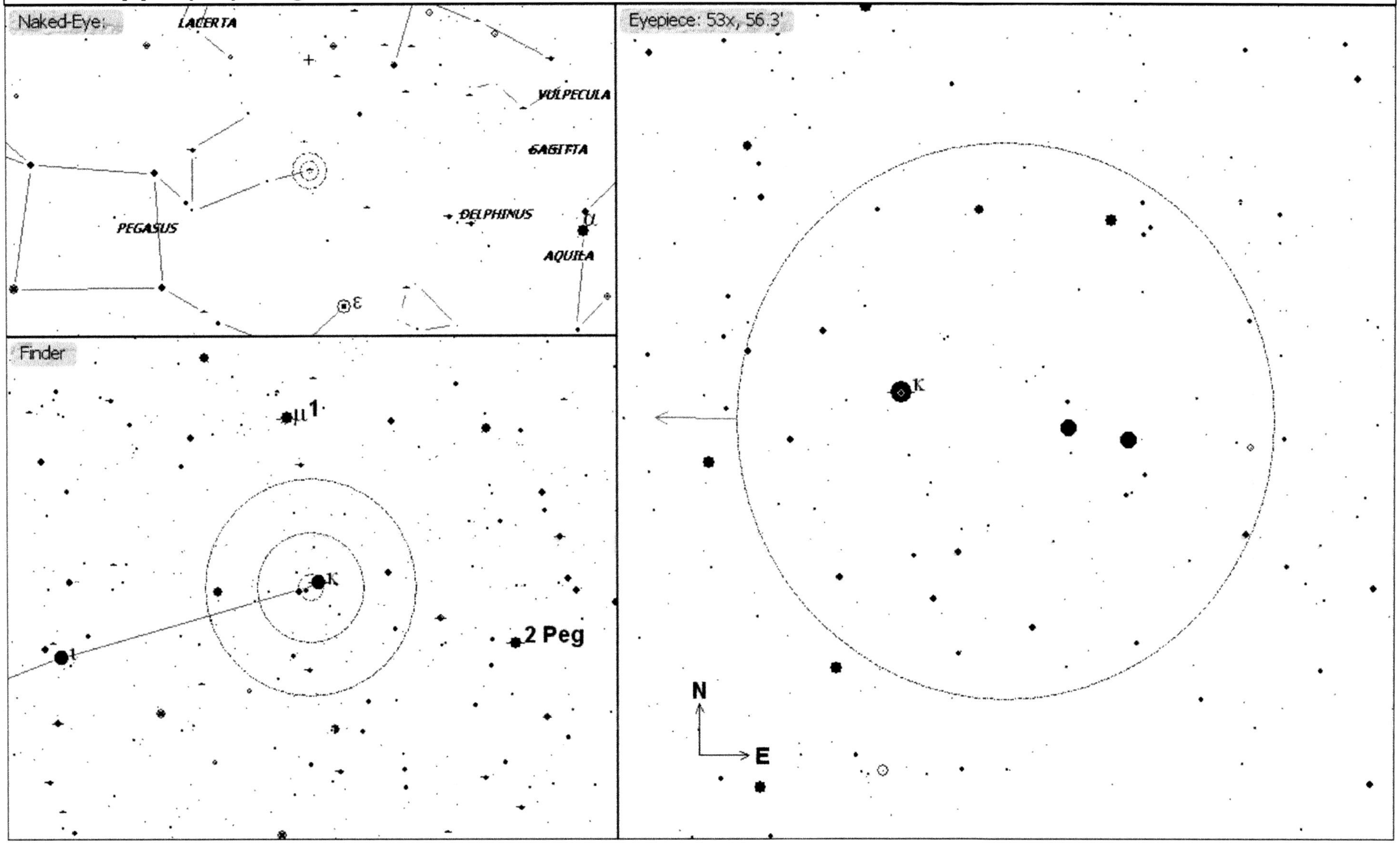

Eyepiece: 53x, 56.3

Naked-Eye

Finder

80mm f/6 Refractor

137

Name	**The 12 Pegasi Asterism**
Location	RA: 21h 46m 04s, Dec: +22° 56' 56" (2000) in Pegasus
Colorful Stars	12 Peg
Found With	8 x 50 finderscope
Description	This asterism is conspicuous in the finder or binoculars. A pair and trio of stars flank 12 Peg. The distance from SAO 89953 to 89982 is 22'.

Principal Stars	
SAO 89952 Single Star aka HD 206966, PPM 113520, HIP 107394 Magnitude: 8.28, A0 Distance: 720 +/- 170.0 ly **SAO 89953** Single Star aka PPM 113519, BD +22 04468 Magnitude: 8.30, G0 ***12 Peg** Single Star aka HR 8321, HD 207089, SAO 89972 Magnitude: 5.29, G8II Distance: 1100 +/- 300.0 ly	**SAO 89975** Single Star aka HD 207135, PPM 113555, HIP 107498 Magnitude: 7.62, A3 Distance: 430 +/- 54.0 ly **TYC 02202-1498 1** Single Star aka SI 204567 Magnitude: 9.62 **SAO 89982** Single Star aka HD 207200, PPM 113565 Magnitude: 874, A0

Object data from SKYTOOLS™ 2 — used with permission.

The 12 Pegasi Asterism / Pegasus / RA: 21h 46m 04s, Dec: +22° 56' 56"

Eyepiece: 60x, 50.0'

Naked-Eye

Finder

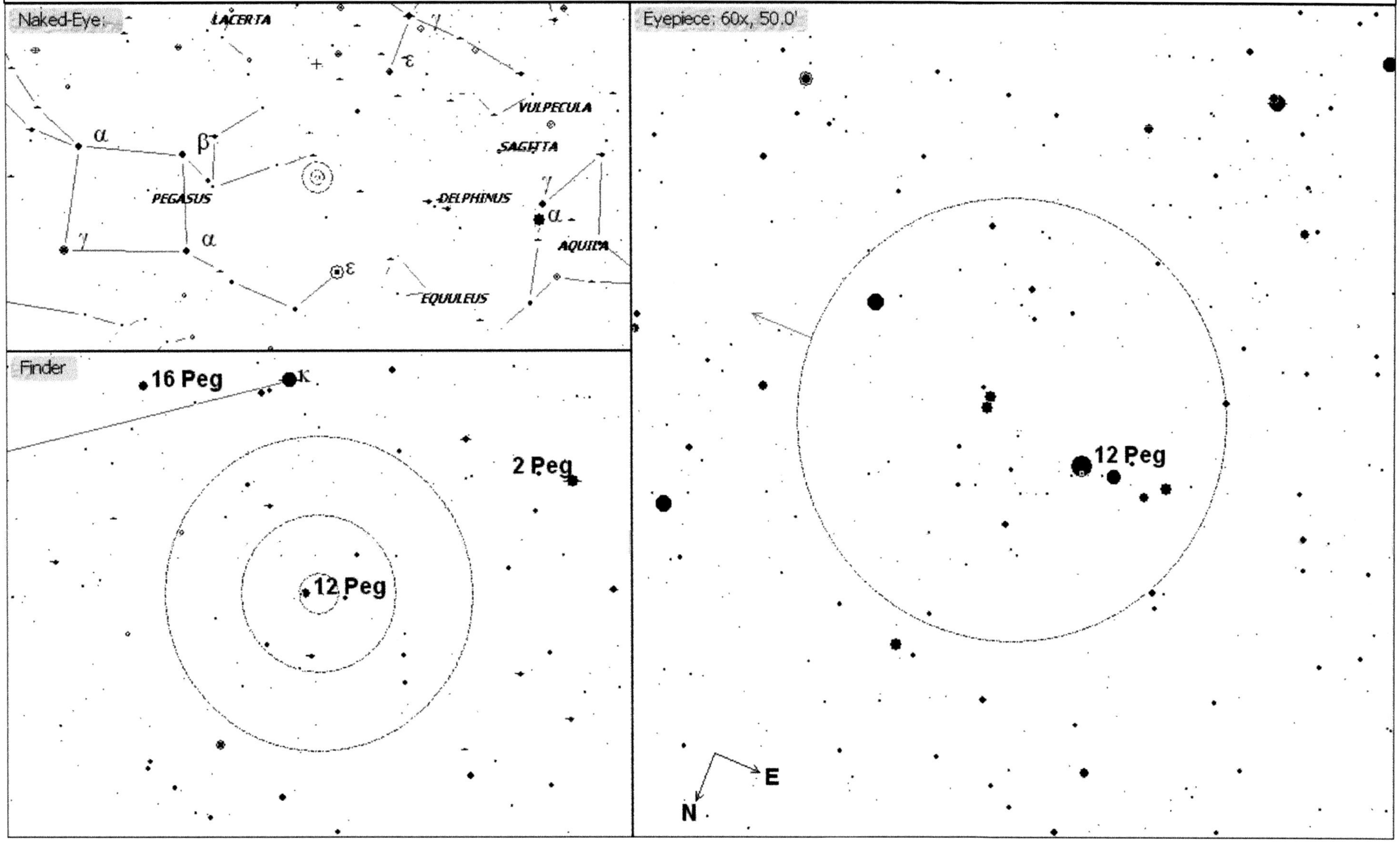

6" f/8 Reflector

Name	**The Little Water Jar Asterism**
Location	RA: 23h 15m 58s, Dec: -01° 25' 26" (2000) in Pisces
Colorful Stars	SAO 146600, 146609
Found With	8x50 Finder
Description	Like the famous original in nearby Aquarius, The Little Water Jar in Pisces empties out its contents southward into the sky. While the Aquarius Water Jar is visible to the eye, binoculars are needed to reveal this one. The distance between SAO 146600 and 146632 is about 50'. Compare with the Pisces "Y" asterism just north of the Circlet of Pisces.

Principal Stars

Water Jar

*SAO 146600 Single Star
aka HD 219461, PPM 181598, HIP 114864
Magnitude: 6.93, K0
Distance: 1100 +/- 340.0 ly
Struve 2995 Multiple Star System
aka HD 219542, HIP 114914, ADS 16642
Magnitude: 7.60
Distance: 180 +/- 19.0 ly
AB: 7.6+8.69 mag, B=SAO 146605
In 1991: PA 32° Sep 5.28"
SAO 146609 Single Star
aka HD 219564, PPM 181602, BD -02 05918
Magnitude: 7.79, K0
SAO 146617 Multiple Star System
aka HD 219657, PPM 181605, HIP 115012, ADS 16649, BU 79
Magnitude: 7.88, G5, Distance: 180 +/- 18.0 ly
AB: 7.88+9.64 mag, B=HIP 115012
Preliminary Orbit: P=388.4 yr, a=1.99"
PA 14° Sep 1.48" (2005.4)

3 Drops

SAO 146627 Single Star
aka HD 219739, PPM 181607
Magnitude: 8.22, G5
SAO 146631 Single Star
aka HD 219771, PPM 207402, HIP 115074
Magnitude: 8.57, G0
Distance: 260 +/- 24.0 ly
SAO 146632 Single Star
aka HD 219788, PPM 207403, HIP 115091
Magnitude: 9.30, F2
Distance: 970 +/- 410.0 ly

Object data from SKYTOOLS™ 2 — used with permission.

The Little Water Jar Asterism / Pisces / SAO 146600 / RA: 23h 15m 58s, Dec: -01° 25' 26"

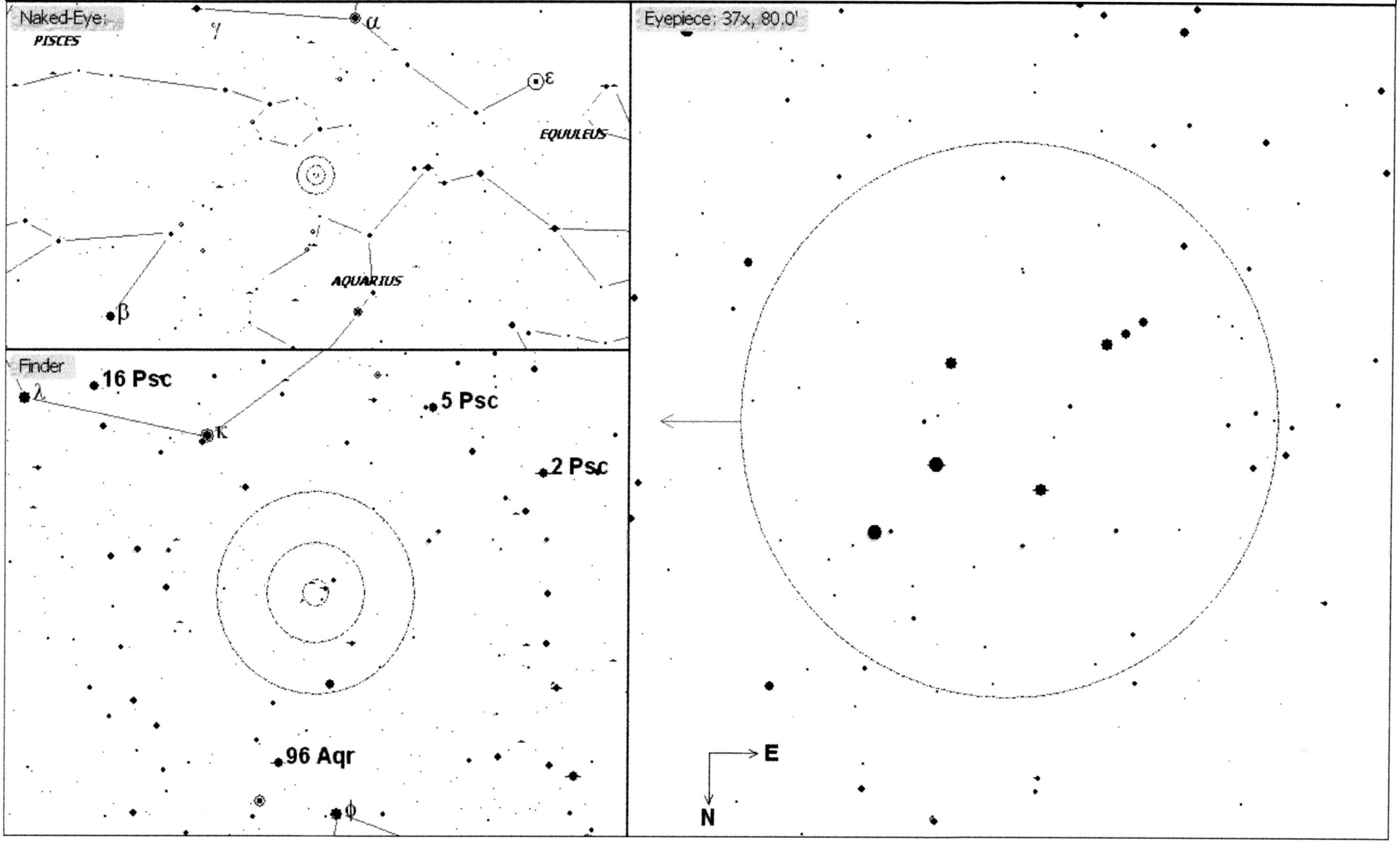

Eyepiece: 37x, 80.0

Naked-Eye
PISCES
EQUULEUS
AQUARIUS

Finder
16 Psc
5 Psc
2 Psc
96 Aqr

6" f/8 Reflector

Name	**The Pisces Y Asterism**
Location	RA: 23h 40m 42s, Dec: +07° 56' 23" (2000) in Pisces
Colorful Stars	SAO 128317, 128320, 128318
Found With	8x50 Finder
Description	A long Y shape, as seen in the finderscope. This asterism lies just north of the Circlet of Pisces, compare it with the Little Water Jar just south of the Circlet. The distance from SAO 128316 to 128317 is 11' 27".

Principal Stars

SAO 128316   Single Star
aka HD 222428, PPM 174321, HIP 116816
Magnitude: 8.82, F5
Distance: 700 +/- 190.0 ly
SAO 128318 Single Star
aka HD 222454, PPM 174324,
Magnitude: 8.25, K2
***SAO 128320**   Single Star
aka HD 222453, PPM 174328,
Magnitude: 8.29, K0

SAO 128324 Multiple Star System
aka HD 222475, PPM 174334,
BU 724, ADS 16924
Magnitude: 9.72, F8
AB: 9.73+10.5 mag
In 1956: PA 95° Sep 0.60"
SAO 128317 Single Star
aka HD 222442, PPM 174322
Magnitude: 8.42, K0

Object data from SKYTOOLS™ 2 — used with permission.

The Pisces Y Asterism / Pisces / SAO 128320 / RA: 23h 40m 42s, Dec: +07° 56' 23"

Eyepiece: 37x, 80.0

Naked-Eye

Finder

58 Peg
59 Peg
77 Peg
82 Peg
80 Peg
32 Psc
θ Y Psc
19 Psc

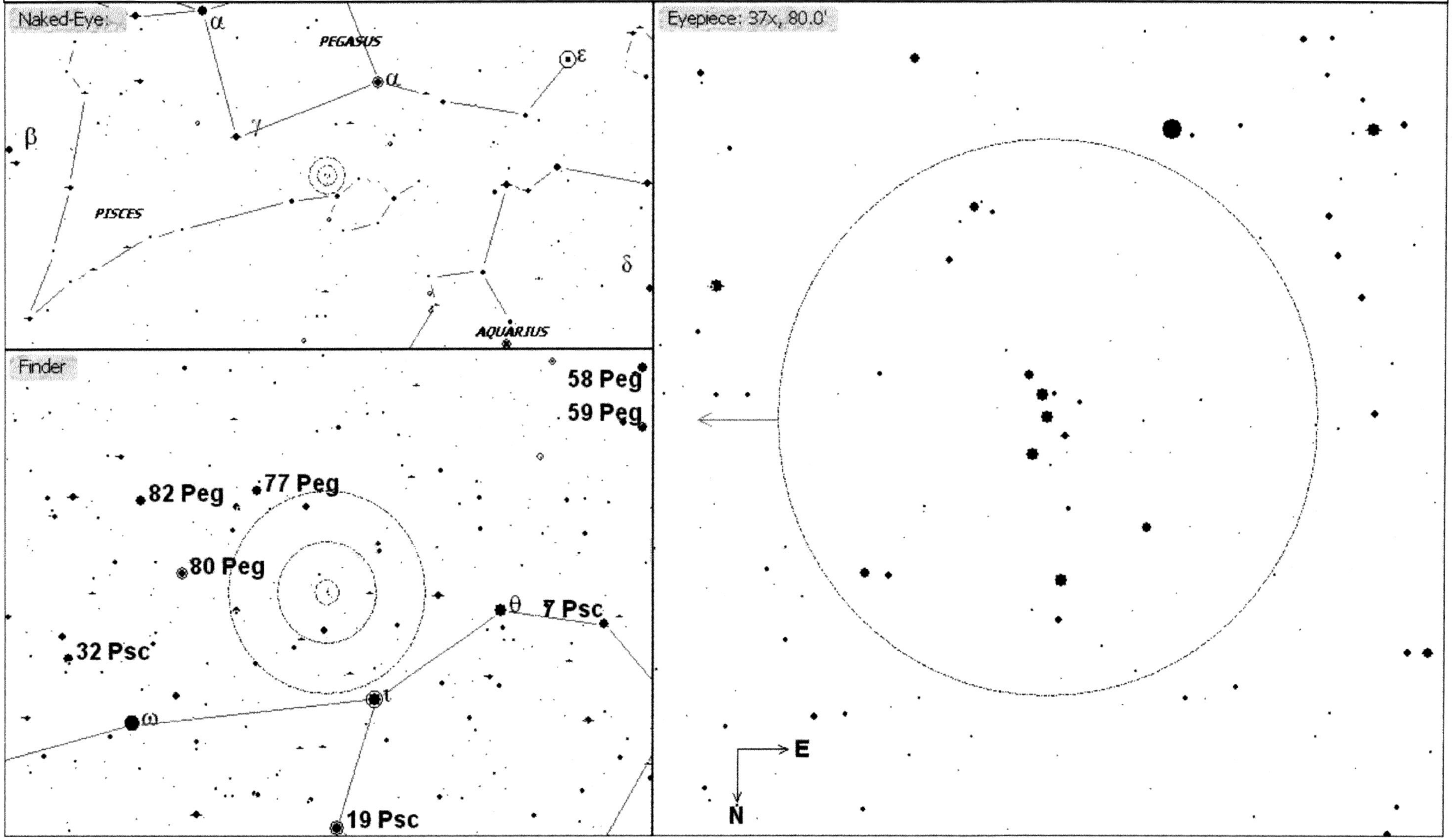

6" f/8 Reflector

143

Name	**The Omega Scorpii Asterism**
Location	RA: 16h 06m 48s, Dec: -20° 40' 09" (2000) in Scorpius
Colorful Stars	Omega ^{2}
Found With	Direct Vision
Description	To the naked eye this pair appears as a fuzzy object in western Scorpius. Careful scrutiny reveals this to be a close pair of stars. A small glass reveals the contrasting colors. The distance between them is 15', one-quarter degree. Nearby are two other wonderful doubles: Beta Sco and Nu Sco. Beta is a Mizar class double, easy in a small scope. Nu is a "Double Double" for medium apertures and high magnifications.

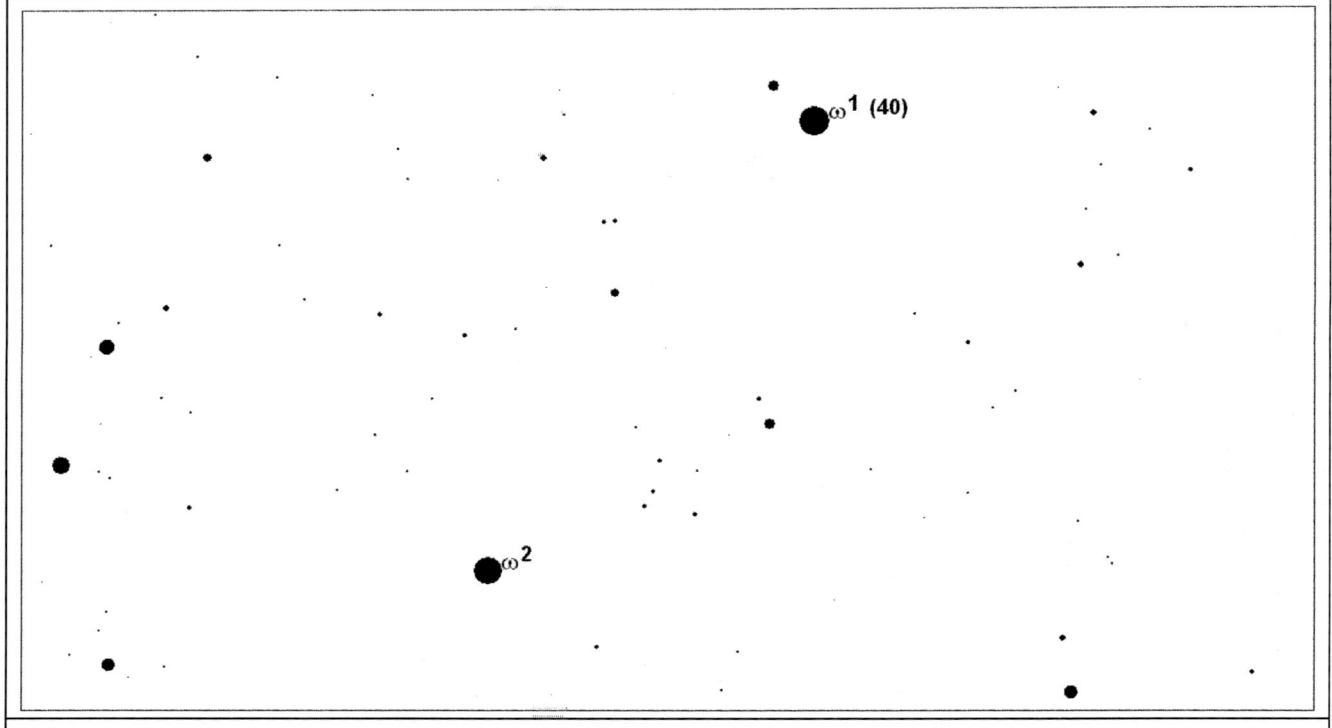

Principal Stars

Omega 1 (ω^1) Sco   Single Star
aka 9 Sco, HR 5993, HD 144470, SAO 184123
Magnitude: 3.97, B1
Distance: 420 +/- 48.0 ly

Omega 2 (ω^2) Sco   Variable Star
aka 10 Sco, HR 5997, HD 144608, SAO 184135, NSV 7457
Suspected Variable
Magnitude: 4.33, G6III
Distance: 260 +/- 19.0 ly

Object data from SKYTOOLS™ 2 — used with permission.

The Omega Scorpii Asterism / Scorpius / Omega¹ (ω^1) Sco / RA: 16h 06m 48s, Dec: -20° 40' 09"

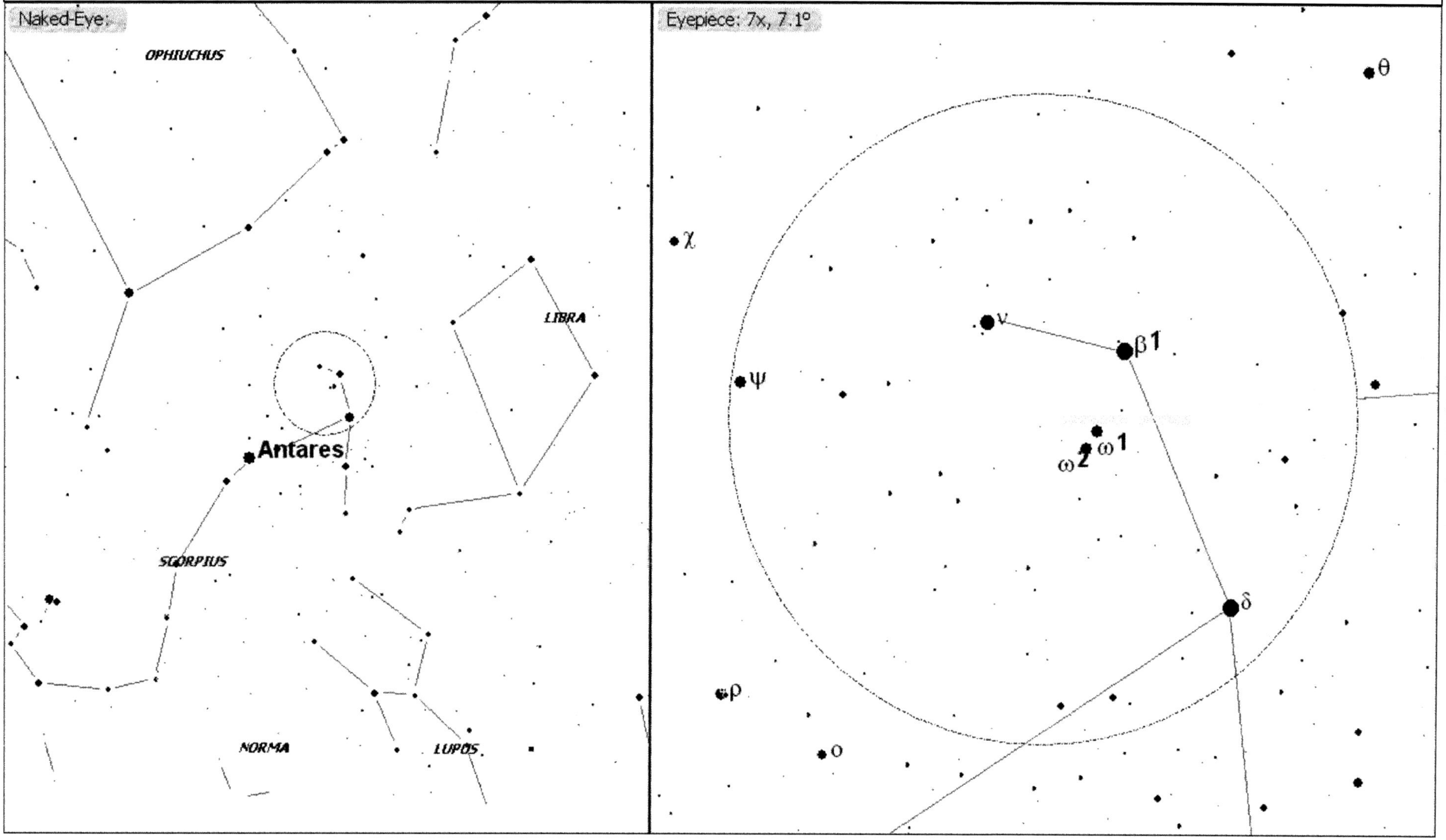

Naked-Eye

Eyepiece: 7x, 7.1°

7 x 50 Binocular

Name	**13 Sagittae Asterism**
Location	RA: 20h 00m 03s, Dec: +17° 30' 59" (2000) in Sagitta
Colorful Stars	13 Sge, SAO 105523
Found With	8 x 50 Finderscope
Description	This asterism is composed of two multiple stars. The brightest member of each is a lovely orange color. It appears as a bright fuzzy object in the finder. The distance between 13 Sge A and SAO 105523 A is 5'. Are they multiple stars or sparse open clusters?

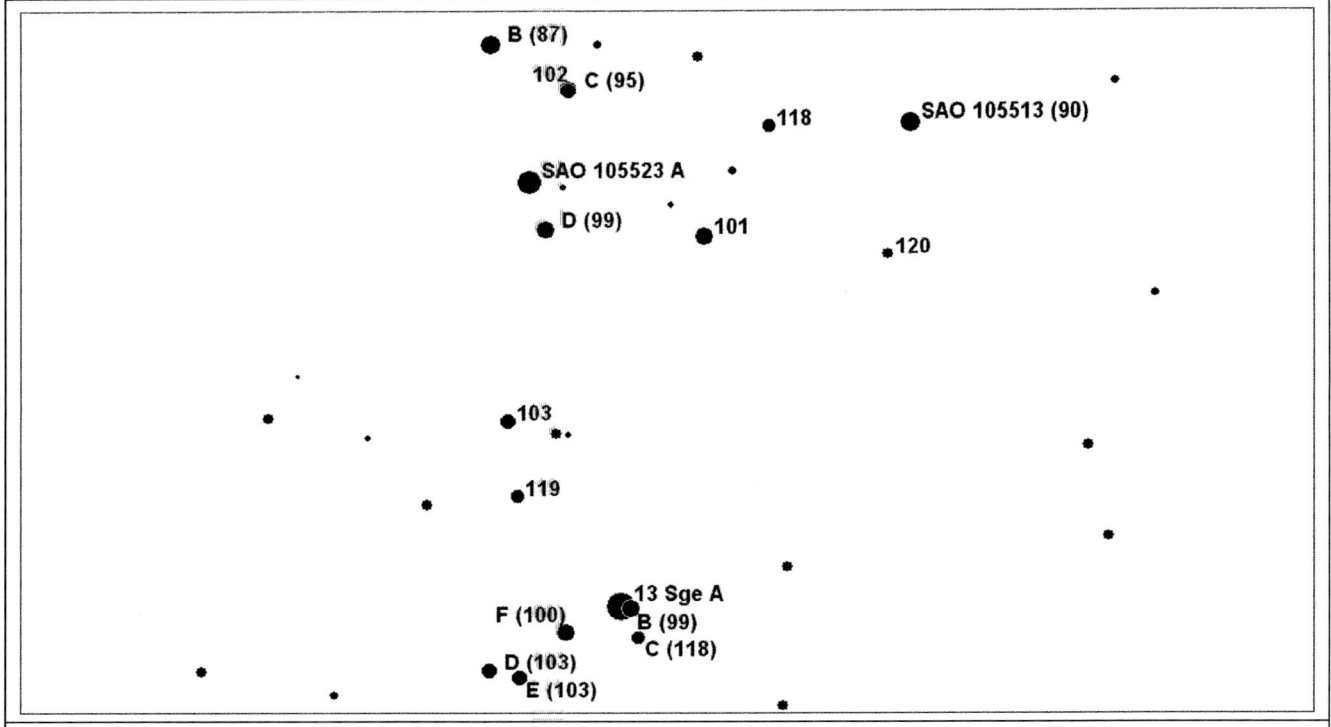

Principal Stars

13 Sge Multiple Star System
aka VZ Sge, GCVS 19870, HR 7645, HD 189577, SAO 105522.
Magnitude: V 5.27 to V 5.57, M0.   Distance: 750 +/- 120.0 ly

AB	5.33+9.9 mag	In 1924: PA 256° Sep 7.50"	
AC	5.33+11.8 mag	In 1934: PA 208° Sep 28.60	ADS 13230
AD	5.33+10.3 mag	In 1904: PA 117° Sep 114.00	D=HD 351107
DE	10.3+10.3 mag	In 1904: PA 256° Sep 23.80"	E=PPM 137389
AF	5.33+10 mag	In 1924: PA 116° Sep 47.10"	ADS 13210

SAO 105523 Multiple Star System
aka HD 189576, PPM 137388, HIP 98443, BD +17 04185.
Magnitude: 6.98, K0, Distance: 1300 +/- 420.0 ly

AB	6.98+8.7 mag	In 1825: PA 15° Sep 114.60"	B=HD 189575
AC	6.98+9.5 mag	In 1879: PA 338° Sep 79.90"	
AD	6.98+9.9 mag	In 1879: PA 198° Sep 39.80"	
AE	6.98+8.8 mag		ADS 13225
EF	8.8+11.8 mag	In 1889: PA 289° Sep 2.30"	ADS 13225

Object data from SKYTOOLS™ 2 — used with permission.

The 13 Sge Asterism / Sagitta / RA: 20h 00m 03s, Dec: +17° 30' 59" (2000)

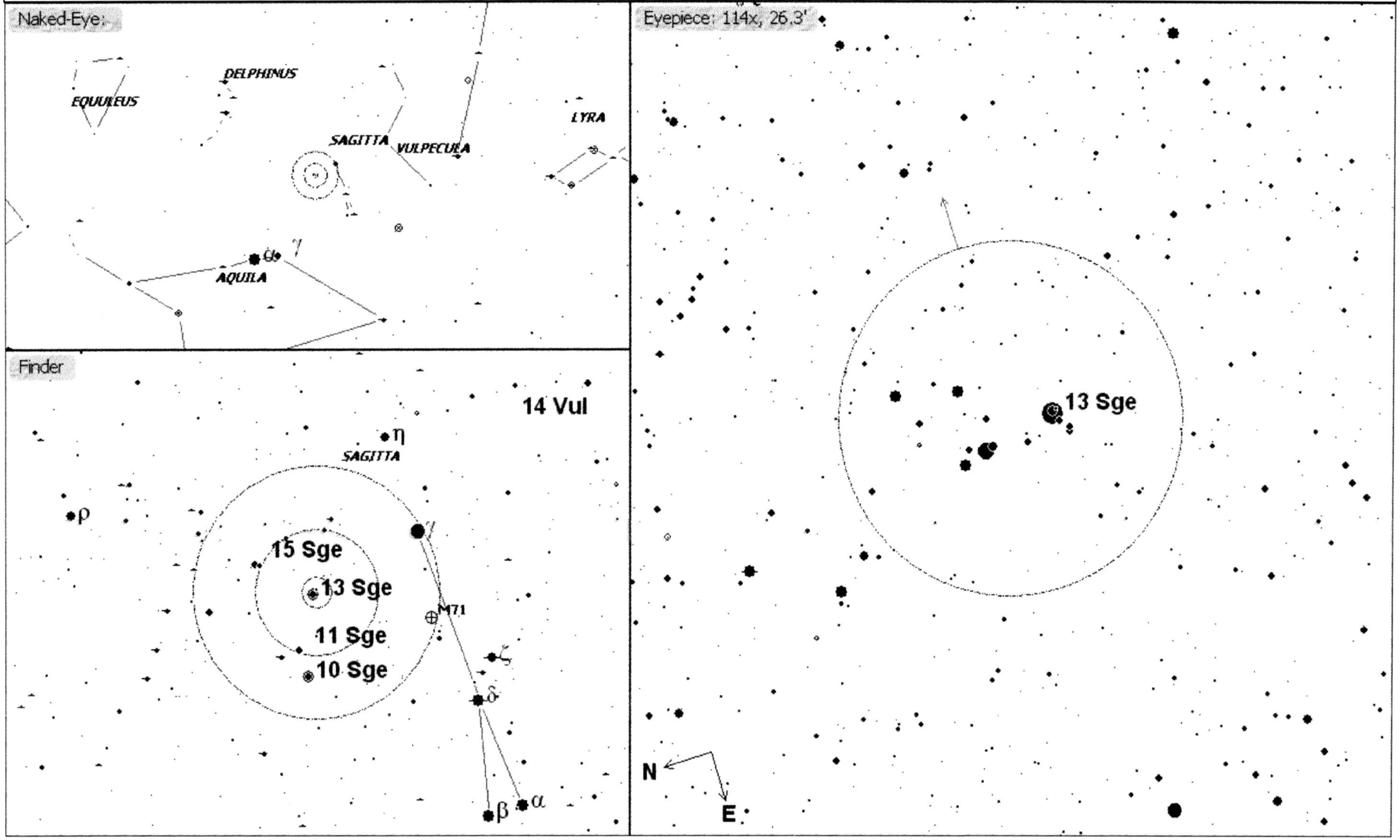

Naked-Eye

Eyepiece: 114x, 26.3'

Finder

6" f/8 Reflector

Name	**15 Sagittae Asterism**
Location	RA: 20h 04m 06s, Dec: +17° 04' 13" (2000) in Sagitta
Found With	8 x 50 Finderscope
Description	This compact asterism is the combination of multiple stars 15 Sge and SAO 105636. At low powers this looks like a small open cluster. The distance between 15 Sge C and SAO 105636 is 7'.

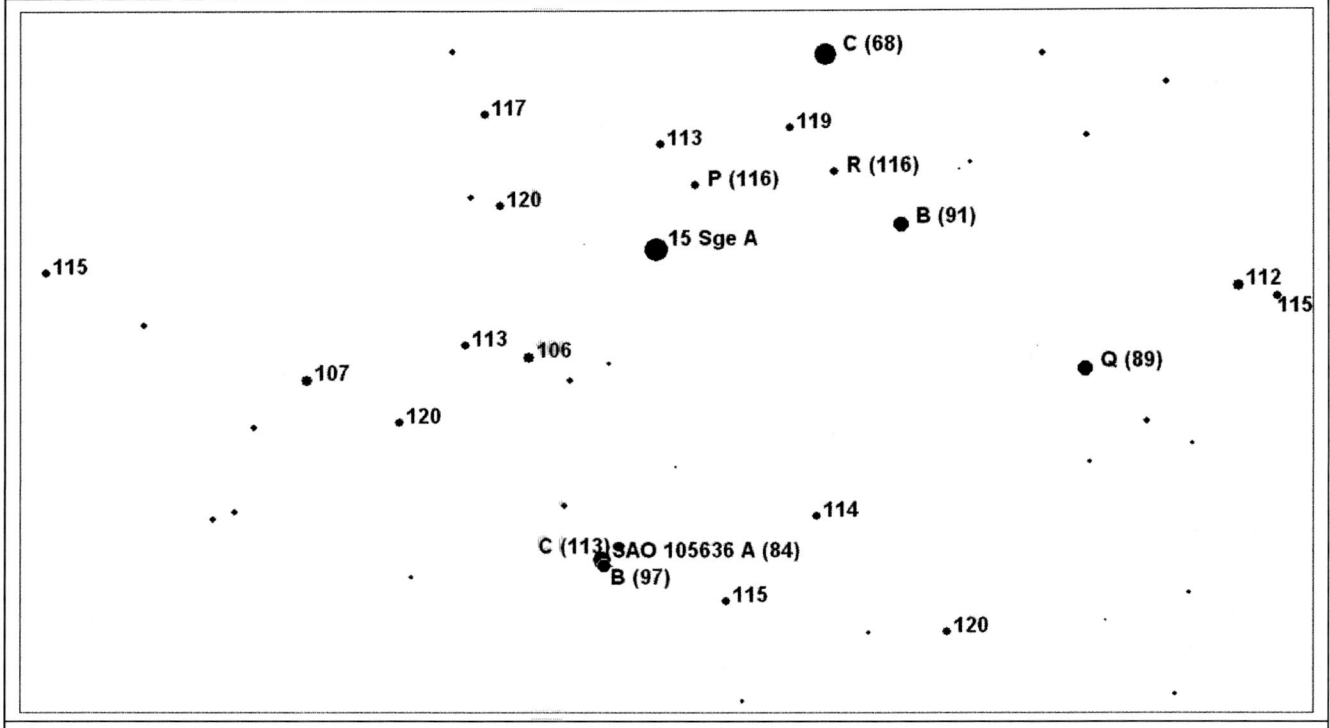

Principal Stars

***15 Sge** Multiple Star System
aka HR 7672, HD 190406, SAO 105635, HIP 98819. Magnitude: 5.80, G1.
Distance: 58 +/- 0.8 ly
Suspected Variable, Type: UNK, Mag: V 5.77 to V 5.80

AB	5.8+9.1 mag	In 1918: PA 276° Sep 190.70"	B=HD 354613
AC	5.8+6.8 mag	In 1918: PA 320° Sep 203.70"	C=HD 190338
AP	5.8+11.6 mag	In 1924: PA 330° Sep 60.00"	
BQ	9.1+8.9 mag	In 1903: PA 231° Sep 183.40"	Q=HD 354614
CR	6.8+11.6 mag	In 1908: PA 184° Sep 93.40	

Struve 2622 Multiple Star System
aka SAO 105636, HD 354615, PPM 137555, HIP 98825. Magnitude: 8.38, F0.
Distance: 850 +/- 480.0 ly

AB	8.38+9.74 mag	In 1991: PA 193° Sep 5.88"	ADS 13323
AC	8.38+11.3 mag	In 1893: PA 308° Sep 17.30"	ADS 13323
BP	9.74+10.4 mag	In 1964: Sep 0.10"	ADS 13323
BQ	9.74+9.5 mag	In 1967: PA 213° Sep 0.20"	ADS 13323

Object data from SKYTOOLS™ 2 — used with permission.

The 15 Sge Asterism / Sagitta / RA: 20h 04m 06s, Dec: +17° 04' 13"

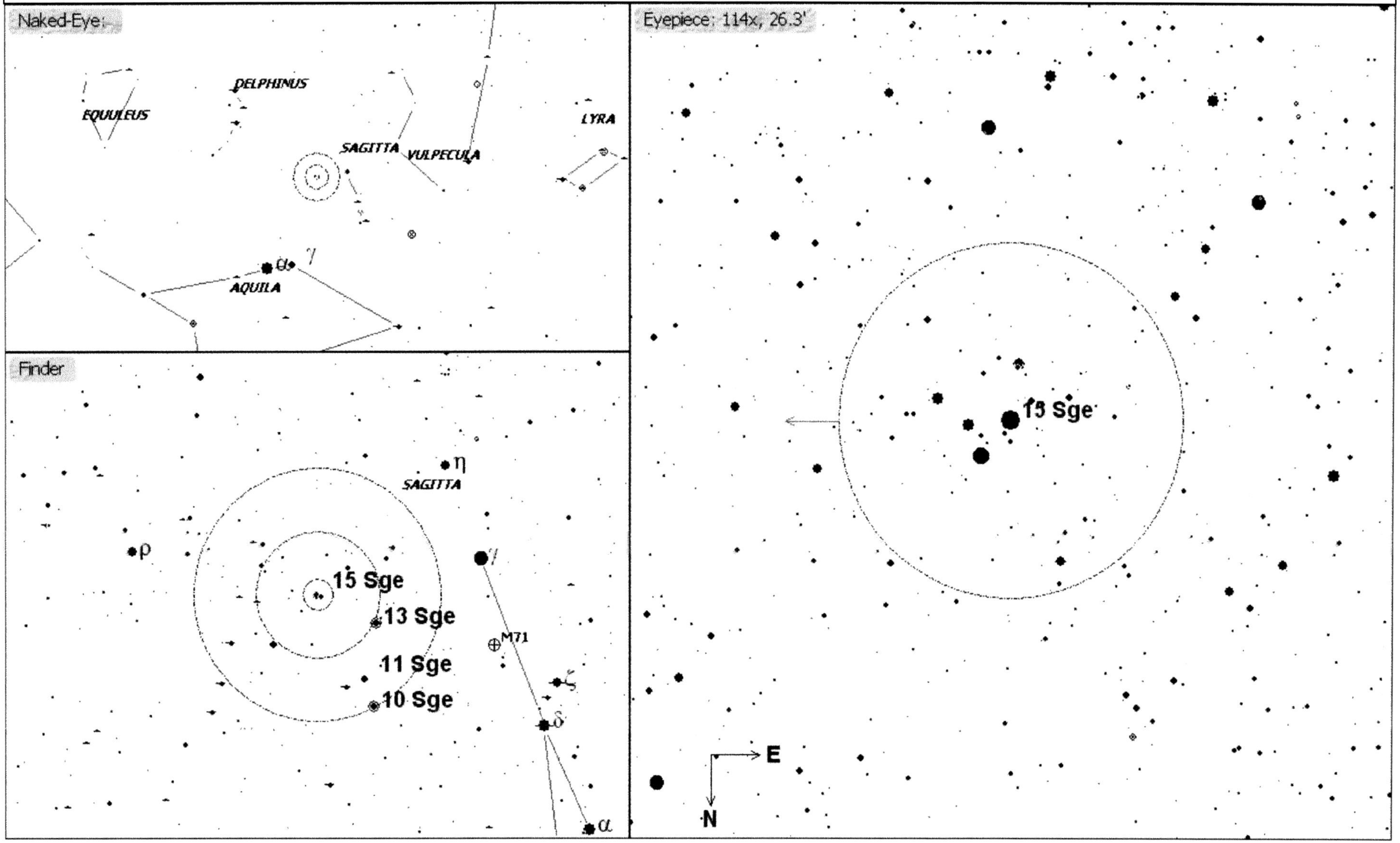

Eyepiece: 114x; 26.3'

Naked-Eye

Finder

6" f/8 Reflector

Name	**The Jewels of the Teapot Asterism**
Location	RA: 18h 28m 06s, Dec: -26° 45' 26" (2000) in Sagittarius
Found With	8 x 50 Finder
Description	A bright trio of double stars just over a degree south of Lambda λ Sgr, the top of the Teapot. SAO 186837 (a true double for large scopes) forms an easy optical double with TYC 06865-2528-1. SAO 186843 (a double for observatory size scopes) is another optical double with SAO 186845. The distance between SAO 186843 and 186863 is 15'.

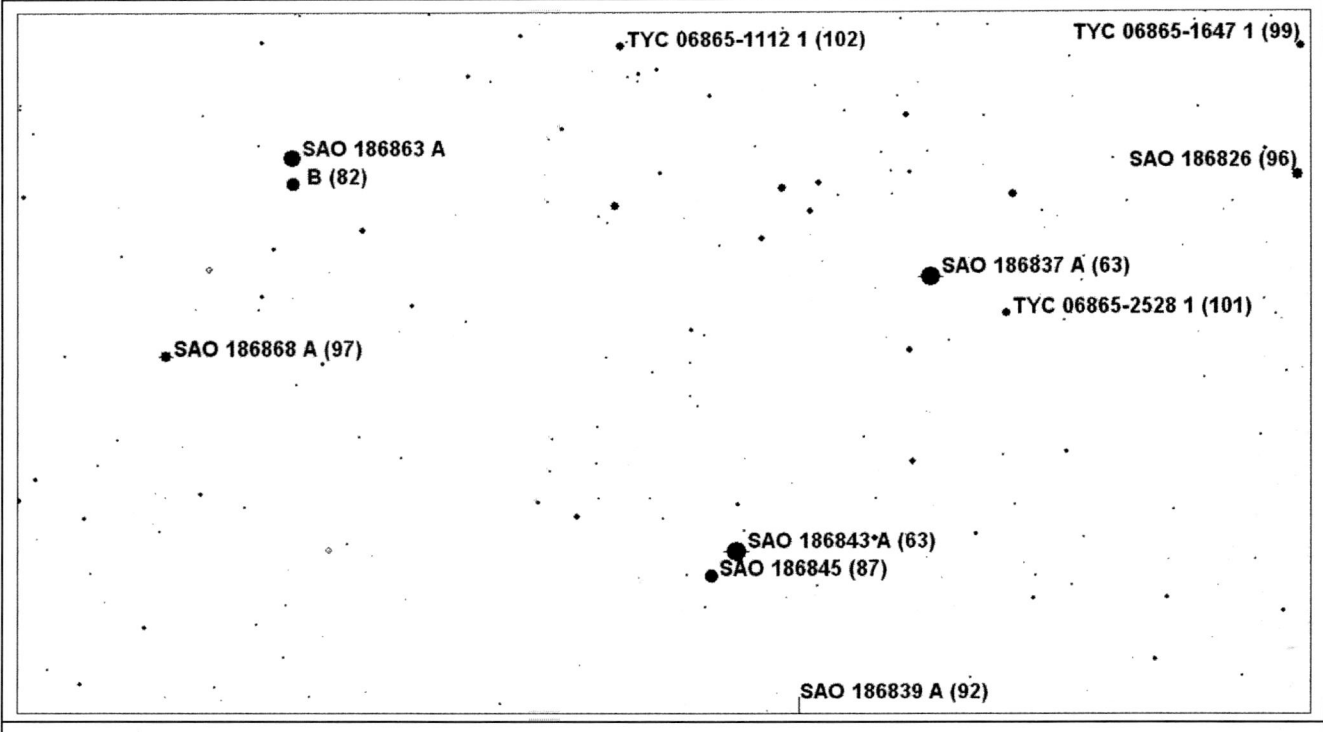

TYC 06865-2528-1 Single Star
aka SI 708304
Magnitude: 10.05

SAO 186837 Multiple Star System
aka HD 169851, HR 6909, PPM 268434,
HIP 90478, BU 133, ADS 11354
Magnitude: 6.31, A7
Distance: 250 +/- 32.0 ly
AB: 6.31+7.2 mag
In 1991: PA 244° Sep 1.01"

*SAO 186843 Multiple Star System
aka HD 169938, HR 6914, PPM 268443,
HIP 90510
Magnitude: 6.28, A3
Distance: 270 +/- 22.0 ly
AB: 6.28+9.31 mag
In 1991: PA 172° Sep 0.33"

SAO 186845 Single Star
aka HD 169969, PPM 268449
Magnitude: 8.70, B9IV

SAO 186863 Multiple Star System
aka HD 170141, HR 6926 , PPM 268467,
HIP 90576, WNO 6
Magnitude: 6.68, A3III
Distance: 400 +/- 50.0 ly
AB: 6.68+8.2 mag, B=HD 170141
In 1890: PA 182° Sep 41.90"

Lambda (λ) Sgr Single Star (Finder Page)
aka 22 Sgr, HR 6913, HD 169916,
SAO 186841
Magnitude: 2.81, K1III
Distance: 77 +/- 1.6 ly

Object data from SKYTOOLS™ 2 — used with permission.

The Jewels of the Teapot Asterism / Sagittarius / SAO 186843 / RA: 18h 28m 06s, Dec: -26° 45' 26"

Finder

Eyepiece: 32x, 1.6°

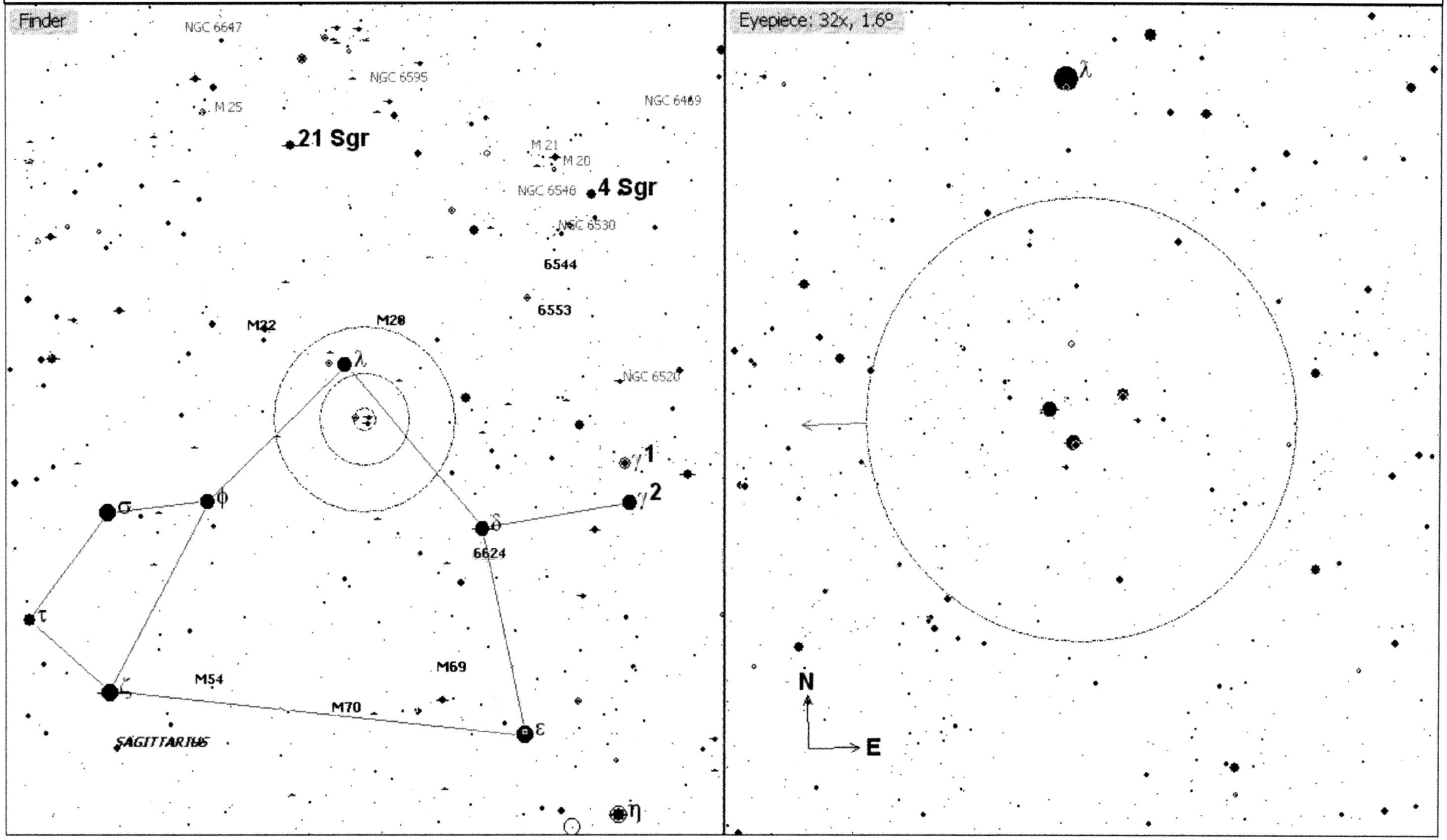

80mm f/6 Refractor

151

Name	**The M22 Asterism**
Location	RA: 18h 33m 53s, Dec: -24° 01' 56" (2000) in Sagittarius
Colorful Stars	SAO 186972, 24 Sgr
Found With	8 x 50 Finder
Description	A pretty group of seven stars one-half degree west of M22. Very conspicuous when star hopping from Lambda λ Sgr to M22.

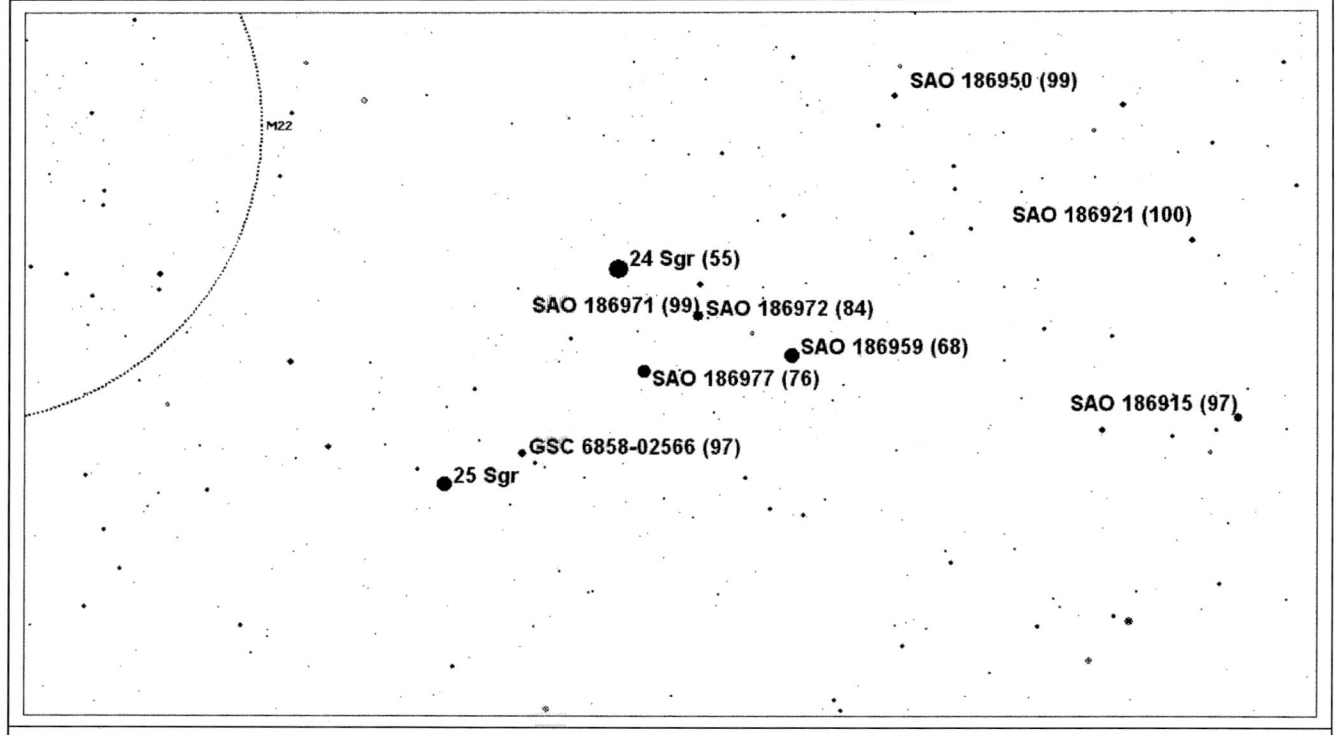

Principal Stars

SAO 186959 Single Star
aka HD 170978, PPM 268563, HIP 90953
Magnitude: 6.81, B3III
Distance: 1100 +/- 320.0 ly

SAO 186971 Single Star
aka PPM 268575, CD -24 14465
Magnitude: 9.86

SAO 186972 Single Star
aka HD 171056, PPM 268576,
TYC 06858-0591 1, SI 707594
Magnitude: 8.38, K1

NGC 6642 Globular Cluster
(see Finder page)
Magnitude: 8.90
Size: 5.8'

SAO 186977 Single Star
aka HD 171097, PPM 268585,
Magnitude: 7.57, B8II

*24 Sgr Single Star
aka HR 6961, HD 171115, SAO 186981
Magnitude: 5.49, K3III

GSC 06858-2566 Single Star
Magnitude: 9.67

25 Sgr Single Star
aka HR 6965, HD 171237, SAO 186995,
PPM 268601, HIP 91066
Magnitude: 6.53, F2II

M 22 Globular Cluster
aka NGC 6656
Magnitude: 5.20
Size: 32.0'
Distance: 9800 ly

Object data from SKYTOOLS™ 2 — used with permission.

The M22 Asterism / Sagittarius / 24 Sgr / RA: 18h 33m 53s, Dec: -24° 01' 56"

Eyepiece: 60x, 50.0'

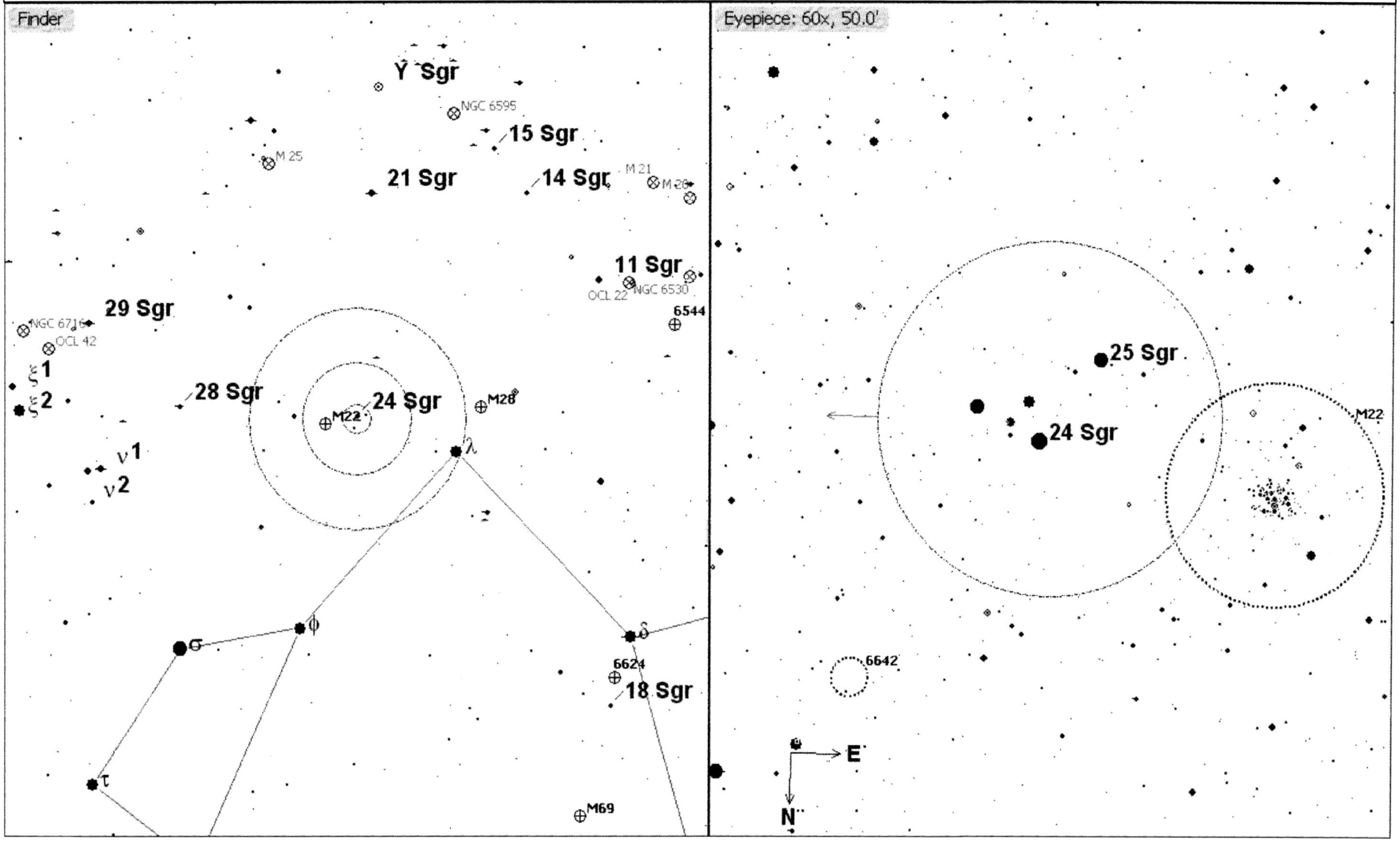

6" f/8 Reflector

Name	**82 Ursae Majoris Asterism**
Location	RA: 13h 39m 30s, Dec: +52° 55' 16" (2000) in Ursa Major
Colorful Stars	SAO 28813, SAO 28816, SAO 28818, SAO 2883, SAO 28842
Found With	8x50 Finderscope
Description	Wonderful grouping of stars around 82 UMa, best seen at low power.

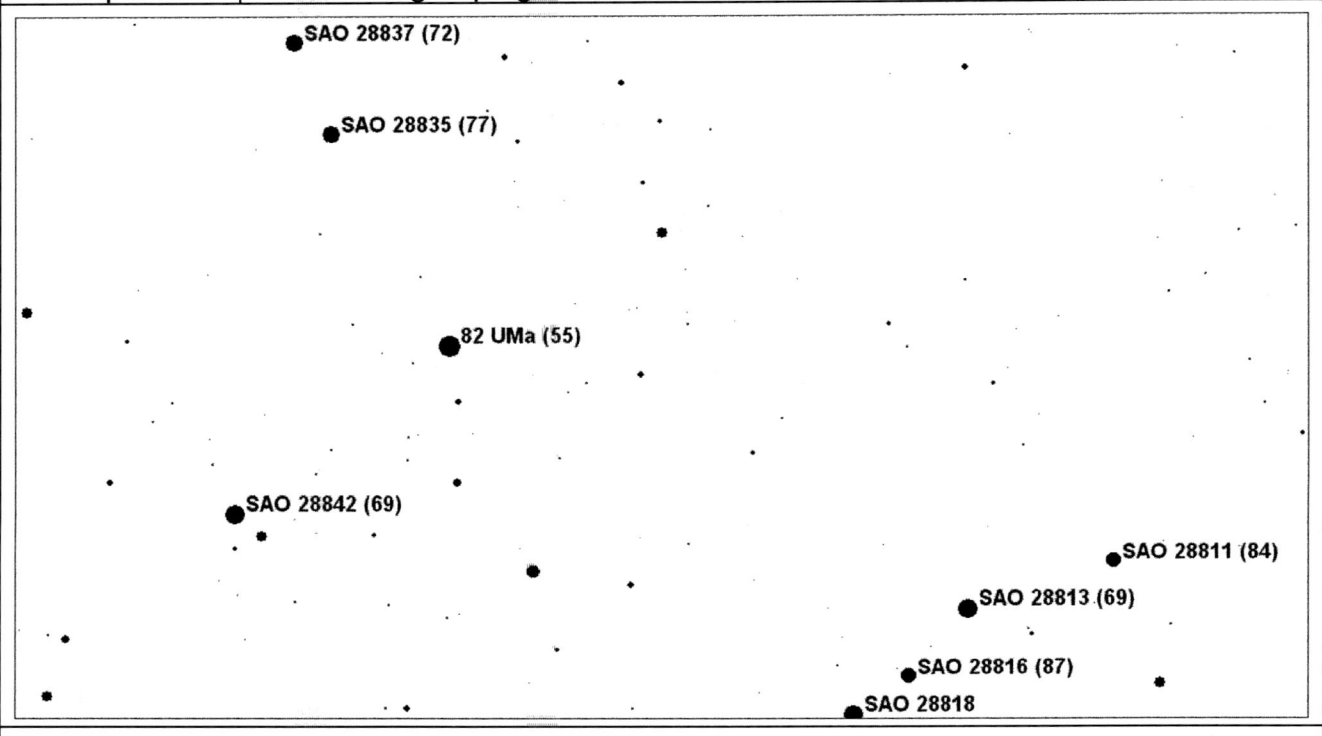

[Star chart showing SAO 28837 (72), SAO 28835 (77), 82 UMa (55), SAO 28842 (69), SAO 28811 (84), SAO 28813 (69), SAO 28816 (87), SAO 28818]

Principal Stars

SAO 28811 Single Star
aka HD 118429
Magnitude: 8.43, F5
Distance: 350 +/- 33.0 ly

SAO 28813 Single Star
aka HD 118575
Magnitude: 6.93, K5
Distance: 560 +/- 67.0 ly

SAO 28816 Single Star
aka HD 234054
Magnitude: 8.71, K2

SAO 28818 Variable Star
aka HD 118668
Magnitude: 6.73, K5
Distance: 1100 +/- 230.0 ly

*82 Uma Single Star
aka SAO 28832, HR 5142, HD 119024
Magnitude: 5.46, A1

SAO 28835 Single Star
aka HD 119146
Magnitude: 7.65, A3
Distance: 500 +/- 63.0 ly

SAO 28837 Single Star
aka HD 119169 , PPM 34140
Magnitude: 7.21, K2
Distance: 420 +/- 43.0 ly

SAO 28839 Single Star
aka HD 234057, PPM 34142
Magnitude: 10.00, G0

SAO 28842 Single Star
aka HD 119214, PPM 34146
Magnitude: 6.88, K0
Distance: 710 +/- 110.0 ly

Object data from SKYTOOLS™ 2 — used with permission.

The 82 Ursae Majoris Asterism / Ursa Major / RA: 13h 39m 30s, Dec: +52° 55' 16"

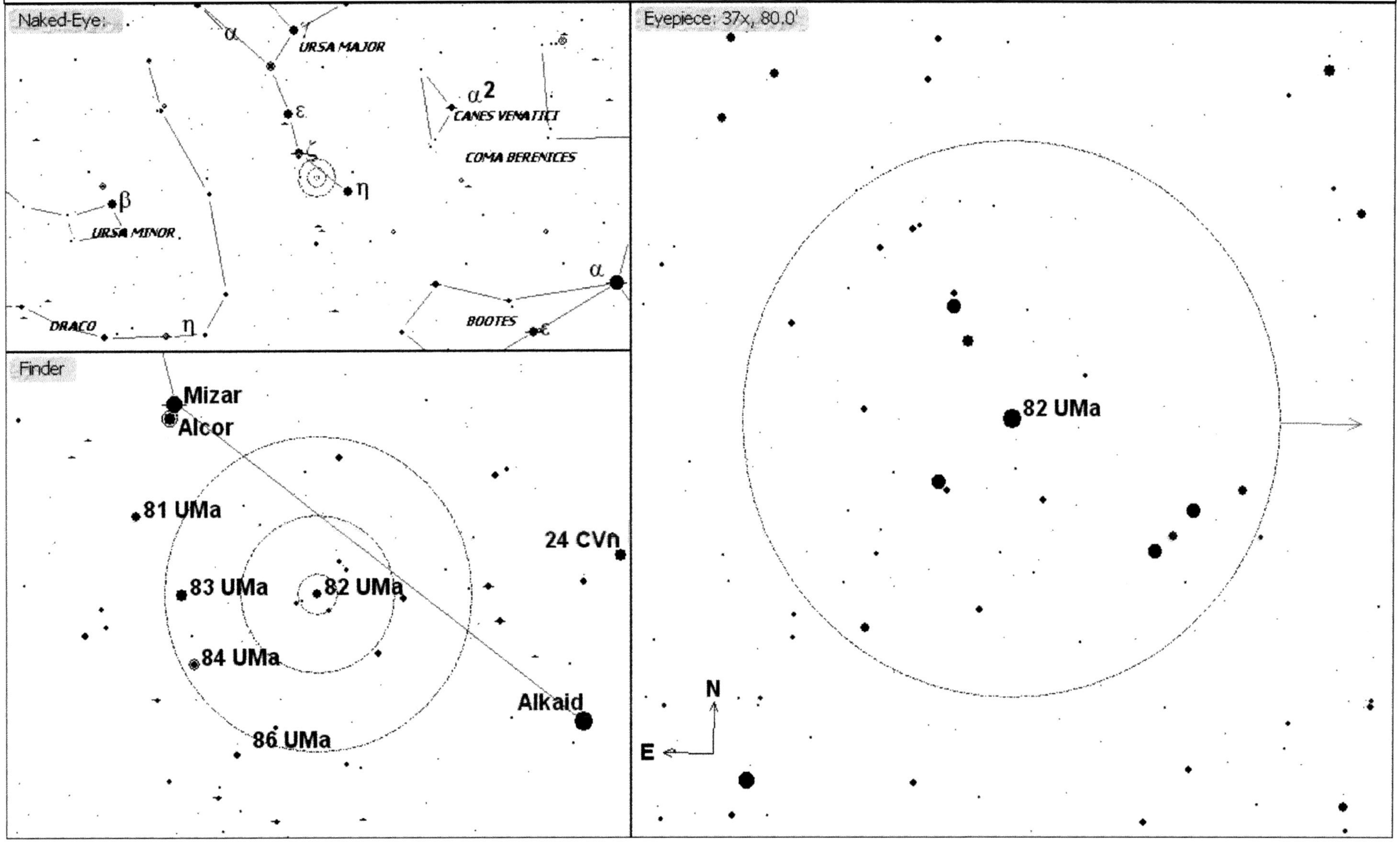

Eyepiece: 37x, 80.0'

Naked-Eye:

Finder:

6" f/8 Reflector

Name	**The Delta (δ) Ursae Minoris Asterism**
Location	RA: 17h 32m 13s, Dec: +86° 35' 11" (2000) in Ursa Minor
Colorful Stars	δ Umi, SAO 2294, 3059
Found With	7 x 50 Finder
Description	A bright circumpolar asterism of four stars.

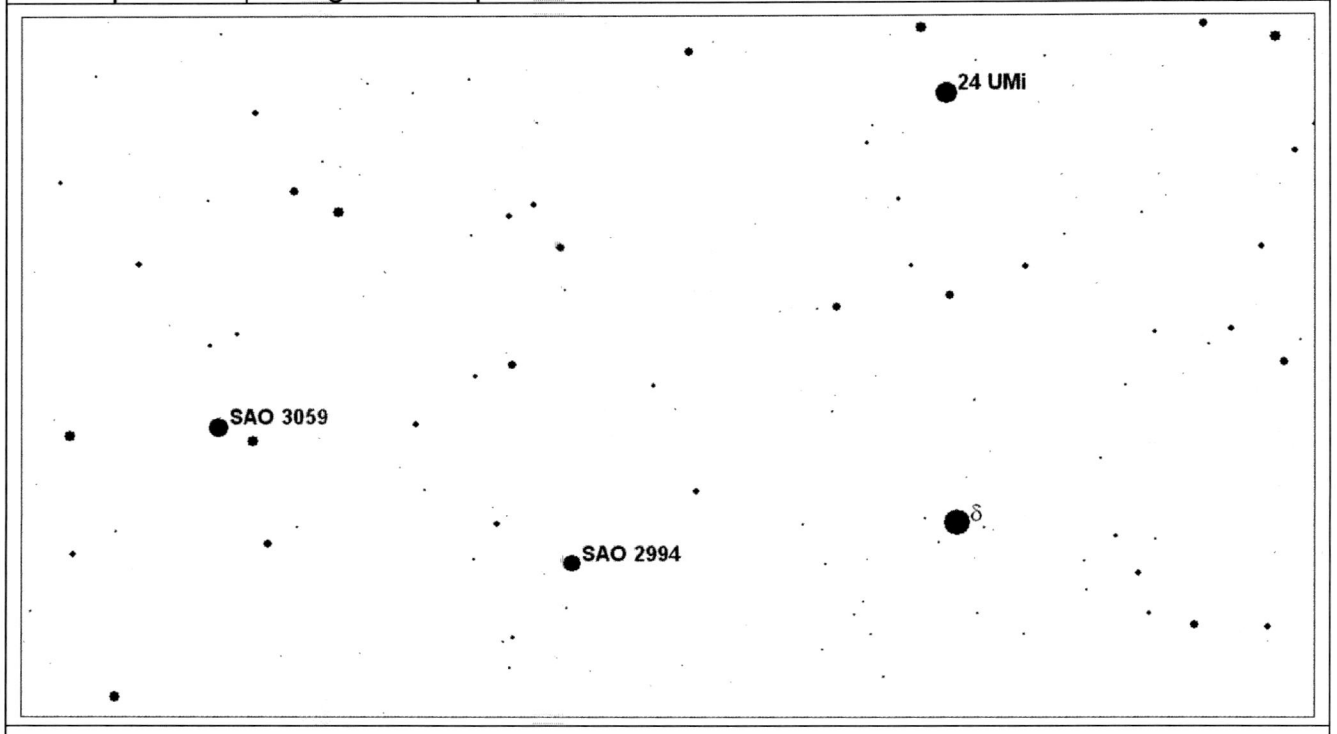

Principal Stars

*Delta Umi Single Star
aka, 23 UMi, HR 6789, HD 166205, SAO 2937
Magnitude: 4.36, K2III
Distance: 180 +/- 4.9 ly

24 Umi Single Star
aka HR 6811, HD 166926, SAO 2940
Magnitude: 5.78, A2
Distance: 160 +/- 3.6 ly

SAO 2994 Single Star
aka HD 170791, PPM 3138, HIP 87663
Magnitude: 7.67, K0
Distance: 690 +/- 91.0 ly

SAO 3059 Single Star
aka HD 174878, PPM 3193, HIP 89465
Magnitude: 6.51, M0
Distance: 1200 +/- 220.0 ly

Object data from SKYTOOLS™ 2 — used with permission.

The Delta (δ) Ursae Minoris Asterism / RA: 17h 32m 13s, Dec: +86° 35' 11"

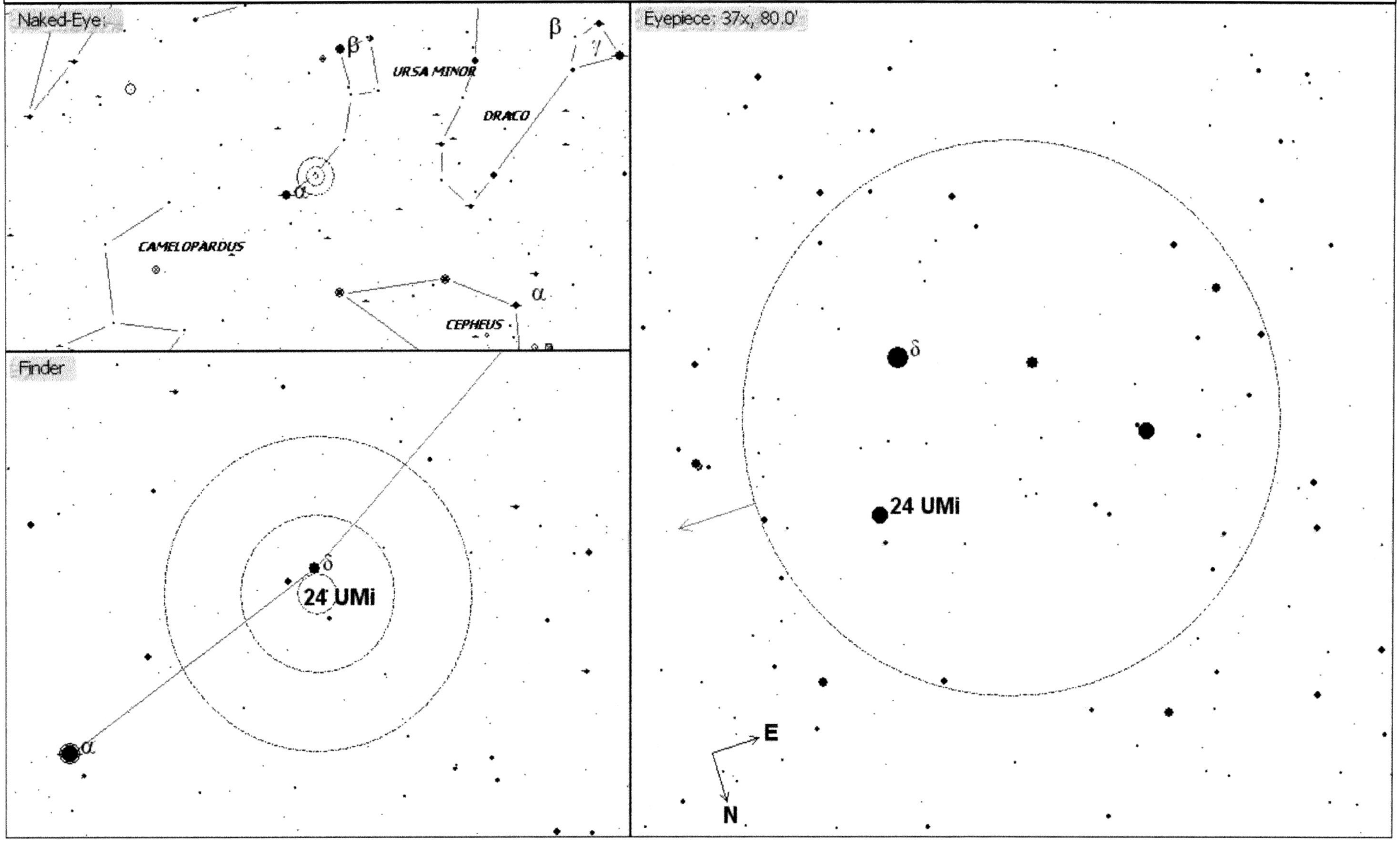

Eyepiece: 37x, 80.0'

24 UMi

δ

E N

Naked-Eye

DRACO
URSA MINOR
β
γ
β
α
CEPHEUS
α
CAMELOPARDUS

Finder

δ
24 UMi
α

6" f/8 Reflector

Name	**The Eta (η) Ursae Minoris Asterism**
Location	RA: 16h 17m 30s, Dec: +75° 45' 19" (2000) in Ursa Minor
Colorful Stars	20 Umi
Found With	7 x 50 Finder
Description	This asterism in the bowl of the Little Dipper is composed of the bright stars η, 19, and 20 UMi. Each star has fainter attendants visible in telescopes, making this a complex asterism.

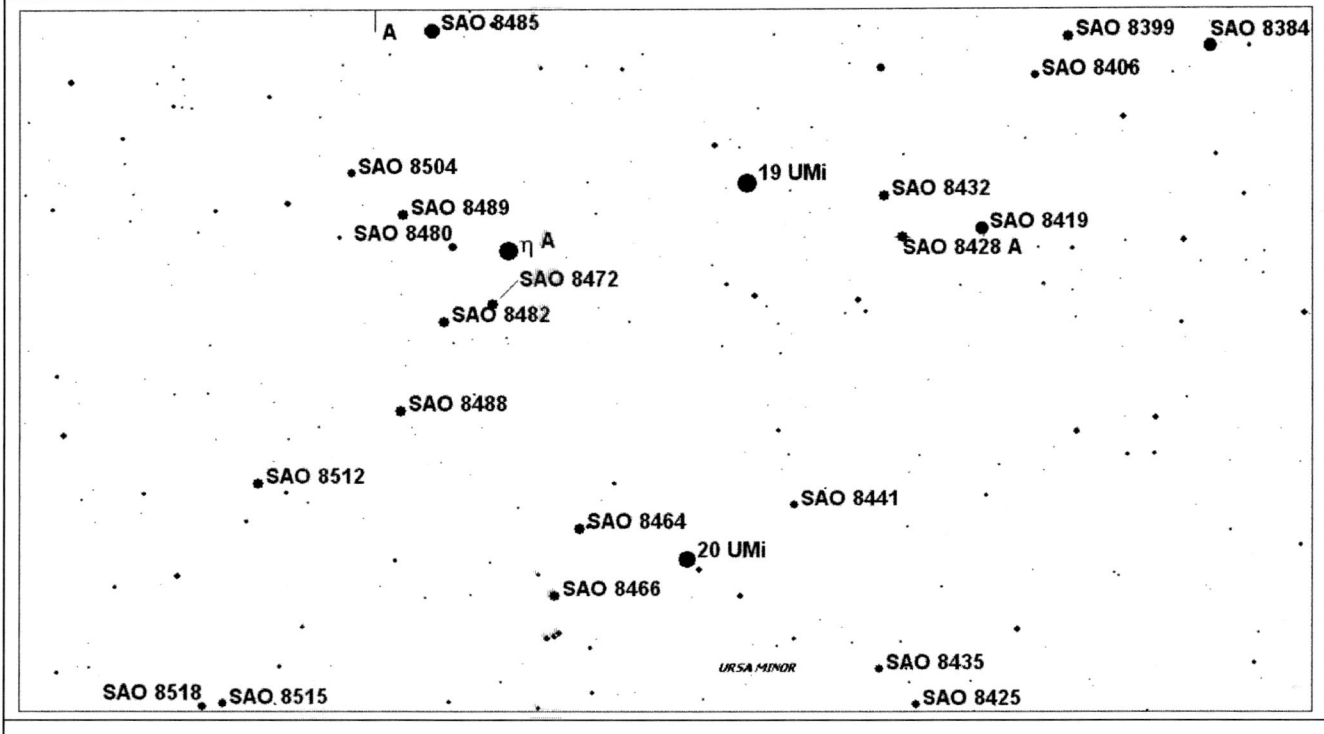

Principal Stars

SAO 8419 Single Star
aka PPM 9097, HIP 78723, BD +76 00590
Magnitude: 7.89

SAO 8428 Multiple Star System
aka PPM 9108, BD +76 00591, ADS 9942
Magnitude: 8.75
AB: 8.78+9.8 mag
In 1904: PA 175° Sep 0.50"

SAO 8432 Single Star
aka PPM 9112, BD +76 00592, TYC 04567-0236 1
Magnitude: 9.08

20 UMi Single Star
aka HR 6082, HD 147142, SAO 8452
Magnitude: 6.36, K2
Distance: 770 +/- 86.0 ly

SAO 8441 Single Star
aka PPM 9124, BD +75 00584, TYC 04567-1630 1
Magnitude: 9.63

19 Umi Single Star
aka HR 6079, HD 146926, SAO 8446
Magnitude: 5.48, B8
Distance: 670 +/- 62.0 ly

SAO 8464 Single Star
aka PPM 9148, BD +75 00587, TYC 04567-1695 1
Magnitude: 9.04

SAO 8466 Single Star
aka PPM 9152, HIP 79714, BD +75 00589
Magnitude: 9.05

*Eta UMi Multiple Star System
aka 21 UMi, HR 6116, HD 148048, SAO 8470, LDS 1844
Magnitude: 4.95, F3
Distance: 97 +/- 1.4 ly
AB: 4.95+16 mag; In 1966: PA 125°Sep 227.00"

SAO 8472 Single Star
aka PPM 9159, BD +76 00597, TYC 04567-1183 1
Magnitude: 9.05

SAO 8480 Single Star
aka PPM 9166, BD +76 00598, TYC 04567-1096 1
Magnitude: 9.75

SAO 8482 Single Star
aka PPM 9167, BD +75 00591, TYC 04567-0978 1
Magnitude: 8.89

SAO 8488 Single Star
aka PPM 9177, BD +75 00592, TYC 04567-1287 1
Magnitude: 8.71

SAO 8489 Single Star
aka PPM 9179, BD +76 00601, TYC 04567-0735 1
Magnitude: 8.26

SAO 8504 Single Star
aka PPM 9190, BD +76 00604, TYC 04567-1081 1
Magnitude: 9.94

SAO 8512 Single Star
aka PPM 9199, BD +75 00595, TYC 04567-1408 1
Magnitude: 8.89

Object data from SKYTOOLS™ 2 — used with permission.

The Eta (η) Ursae Minoris Asterism / RA: 16h 17m 30s, Dec: +75° 45' 19"

Eyepiece: 37x, 80.0

20 UMi
η
19 UMi

E
N

Naked-Eye

α
η
β
β
α
DRACO
URSA MINOR
CAMELOPARDUS

Finder

19 UMi
η
θ
ζ
URSA MINOR

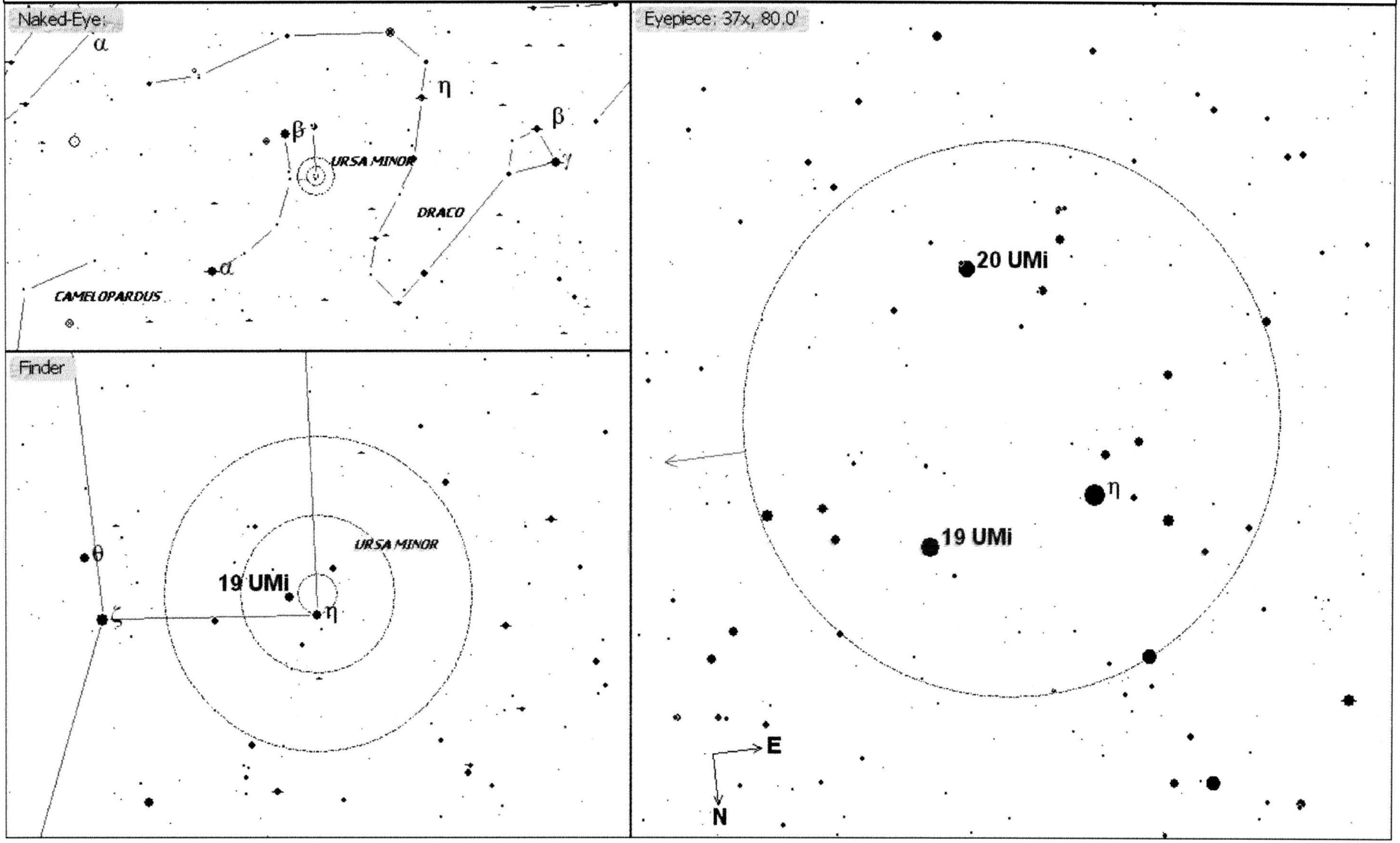

6" f/8 Reflector

Name	**The Ursa Minoris Pseudocluster A**
Location	RA: 16h 30m 39s, Dec: +77° 26' 47" (2000) in Ursa Minor
Colorful Stars	SAO 8597
Found With	7 x 50 finder
Description	This is the southern asterism of a pair located between Eta η and Epsilon ε UMi. Asterism "B" lies one degree to the north. This concentration of stars resembles a sparse open cluster. There are 15 stars brighter than 10th magnitude within a one degree circle, centered on TYC 04571-0615-1. Numerous fainter stars form pairs and chains in the north-east half of the asterism.

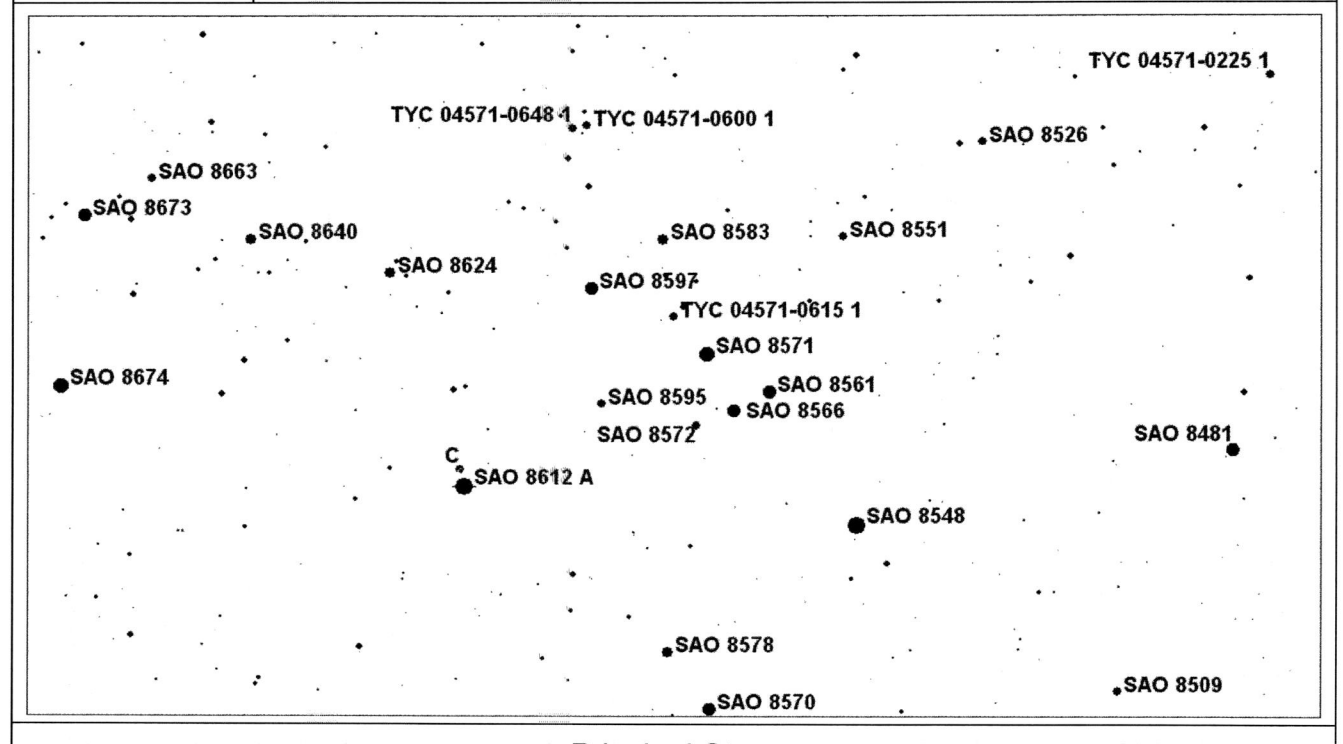

Principal Stars

*SAO 8548 Single Star
aka HD 150275, HR 6191, PPM 9233, HIP 80850
Magnitude: 6.35, G5
Distance: 410 +/- 30.0 ly
SAO 8551 Single Star
aka PPM 9234, BD +78 00560, TYC 04571-1300 1
Magnitude: 9.53
SAO 8561 Single Star
aka HD 150730, PPM 9250, BD +77 00628,
Magnitude: 7.55, A2
SAO 8566 Single Star
aka HD 150902, PPM 9255, HIP 81156, BD +77 00629
Magnitude: 7.93, G5
Distance: 1300 +/- 360.0 ly
SAO 8571 Single Star
aka HD 151043, PPM 9260, HIP 81219, BD +78 00562
Magnitude: 6.89, F8
Distance: 300 +/- 16.0 ly
SAO 8572 Single Star
aka PPM 9263, BD +77 00631, TYC 04571-0737 1
Magnitude: 9.31
TYC 04571-0615-1 Single Star
aka SI 495885
Magnitude: 9.25
SAO 8583 Single Star
aka HD 151341, PPM 9269, BD +78 00564
Magnitude: 8.69, F0

SAO 8595 Single Star
aka PPM 9283, BD +77 00633, TYC 04571-1061 1
Magnitude: 9.26
SAO 8597 Single Star
aka HD 151698, PPM 9284, HIP 81528, BD +78 00565
Magnitude: 8.03, K0
Distance: 1300 +/- 410.0 ly
TYC 04571-0600 1 Single Star
aka SI 495884
Magnitude: 9.36
TYC 04571-0648 1 Single Star
aka SI 495889
Magnitude: 9.80
SAO 8612 Multiple Star System
aka HD 152303, HR 6267, PPM 9301,
HIP 81854, ADS 10214
Magnitude: 5.99, G3
Distance: 120 +/- 2.9 ly
AB: 5.99+10.2 mag, B=HIP 81854
In 1991: PA 178° Sep 2.55"
AC: 5.99+9.8 mag
In 1959: PA 14° Sep 115.00"
SAO 8624 Single Star
aka HD 152810, PPM 9313, HIP 82061, BD +78 00568
Magnitude: 8.47, G5
Distance: 1400 +/- 500.0 ly

Object data from SKYTOOLS™ 2 — used with permission.

The Ursa Minor Pseudocluster 'A' / Ursa Minor / SAO 8548 / RA: 16h 30m 39s, Dec: +77° 26' 47"

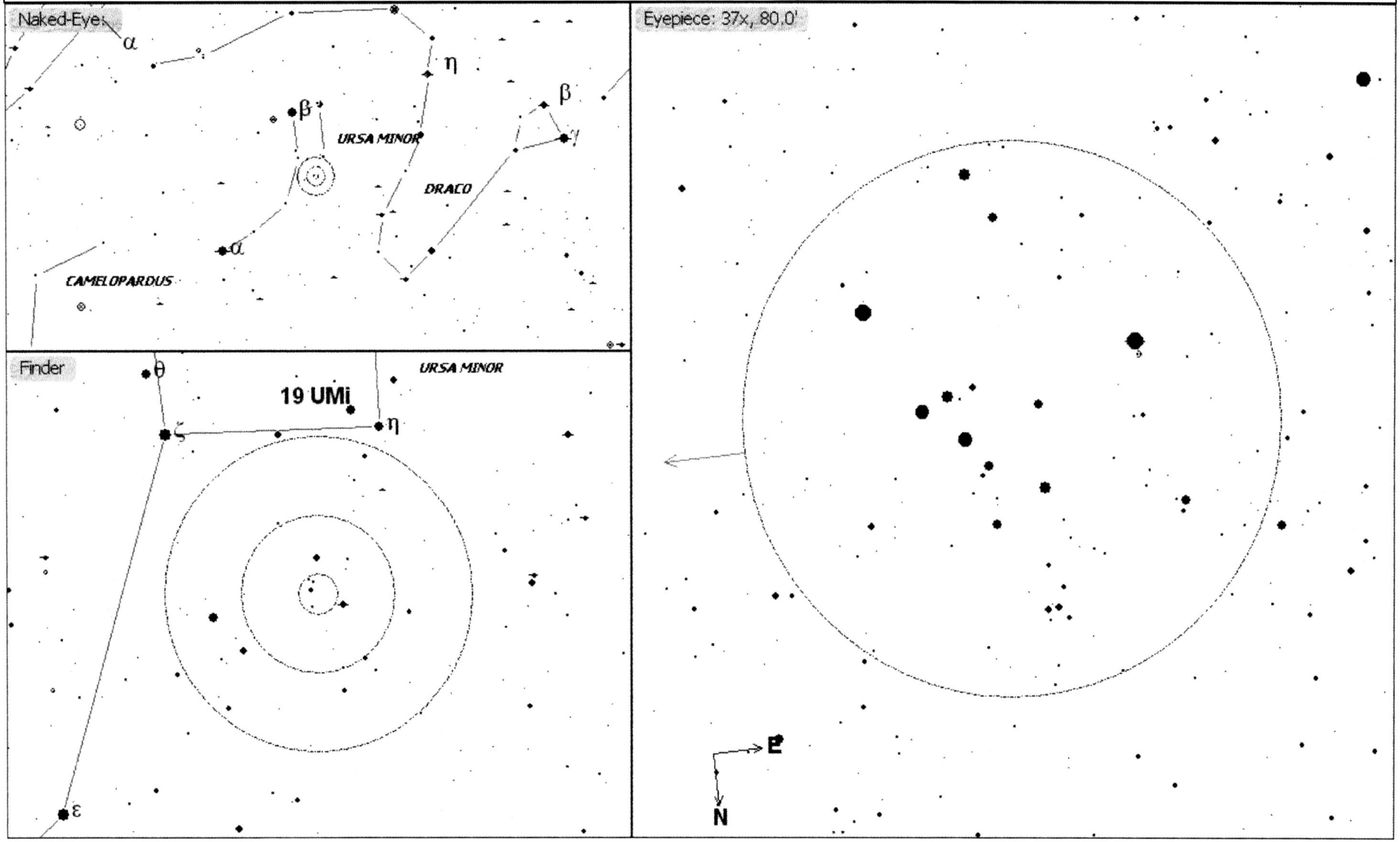

Eyepiece: 37x, 80.0'

Naked Eye

Finder

19 UMi

6" f/8 Reflector

Name	**The Ursa Minoris Pseudocluster B**
Location	RA: 16h 37m 53s, Dec: +78° 55' 07" (2000) in Ursa Minor
Colorful Stars	SAO 8588, 8592
Found With	7 x 50 finder
Description	This asterism resembles a sparse open cluster. It is smaller and fainter than its counterpart located one degree to the south. It is located half way between Eta η and Epsilon ε UMi. Numerous 11th and 12th magnitude stars form groups and chains amidst the brighter SAO stars. It meets three criteria of asterisms: noticeable in a finderscope, distinct from the surrounding field, and resemblance to an open cluster.

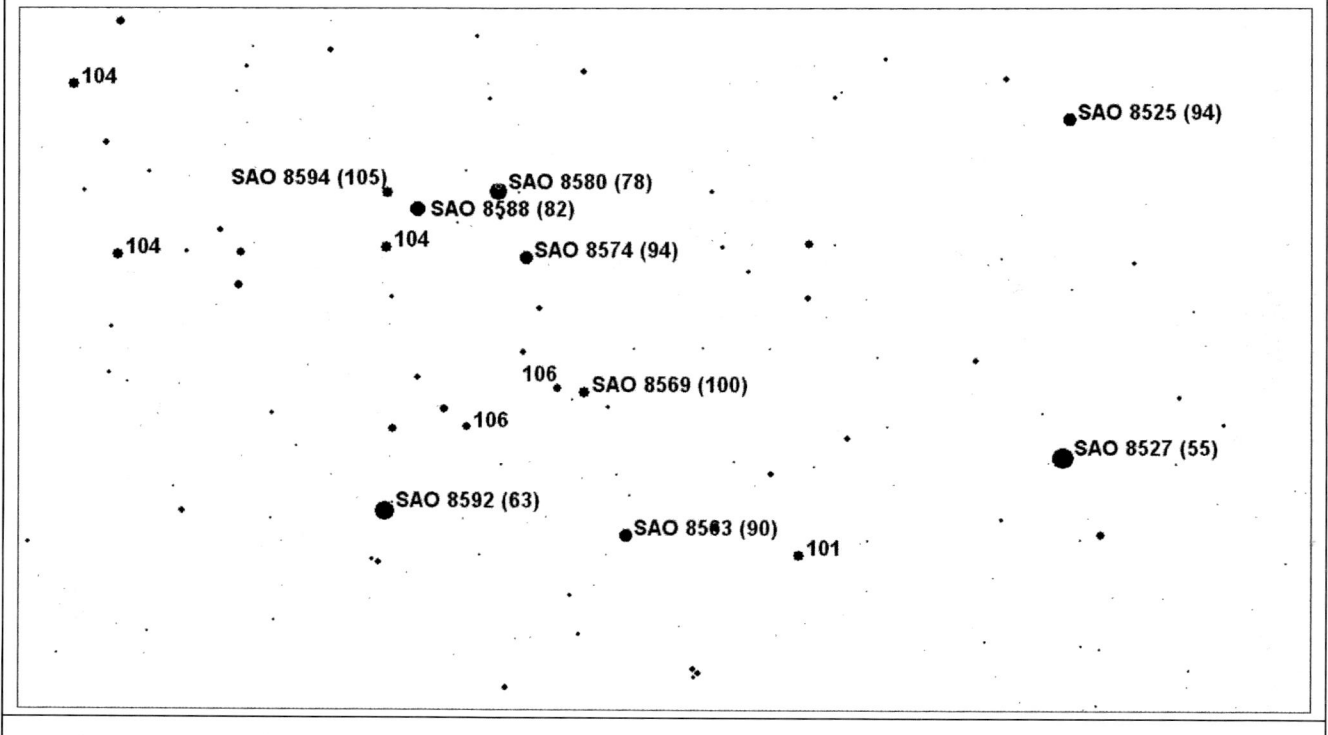

Principal Stars

SAO 8563 Single Star
aka PPM 9252, BD +79 00504
Magnitude: 9.01

SAO 8569 Single Star
aka PPM 9254, BD +79 00506,
TYC 04575-1065 1
Magnitude: 10.03

SAO 8574 Single Star
aka PPM 9261, BD +79 00507,
TYC 04575-0723 1
Magnitude: 9.43

SAO 8580 Single Star
aka HD 151340, PPM 9264, HIP 81270,
BD +79 00508
Magnitude: 7.79, F2
Distance: 370 +/- 27.0 ly

SAO 8588 Single Star
aka HD 151588, PPM 9271, HIP 81396
Magnitude: 8.24, K0
Distance: 960 +/- 200.0 ly

SAO 8594 Single Star
aka PPM 9277, BD +79 00513,
TYC 04575-0565 1
Magnitude: 10.48

*SAO 8592 Single Star
aka HD 151623, HR 6238, PPM 9274,
HIP 81428
Magnitude: 6.33, K0
Distance: 410 +/- 29.0 ly

Object data from SKYTOOLS™ 2 — used with permission.

The Ursa Minor Pseudocluster 'B' / Ursa Minor / SAO 8592 / RA: 16h 37m 53s, Dec: +78° 55' 07"

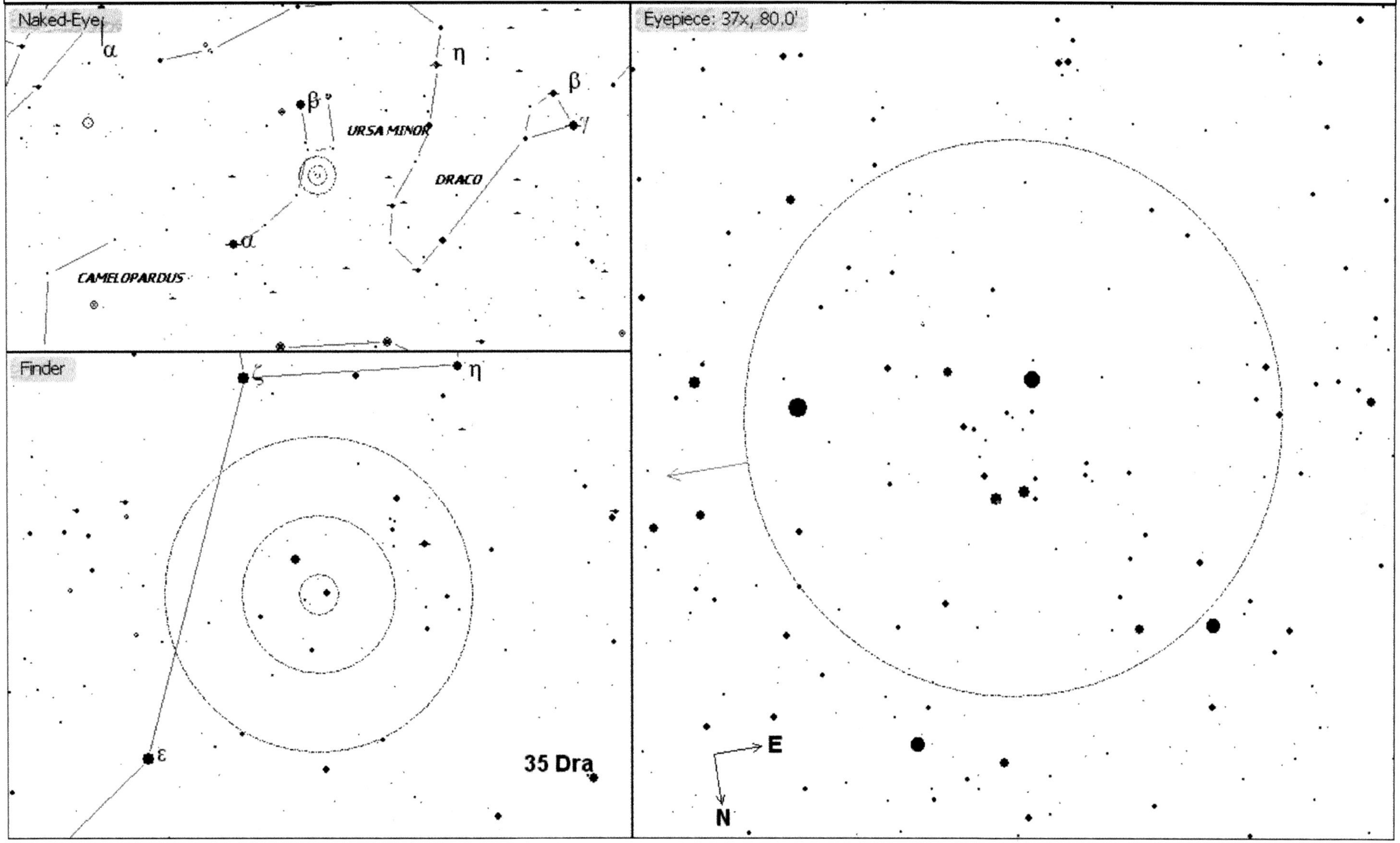

Eyepiece: 37x, 80.0'

Naked-Eye

Finder

35 Dra

6" f/8 Reflector

Name	**The Coronae Virginis Asterism**
Location	RA: 12h 46m 23s, Dec: +09° 32' 23" (2000) in Virgo
Colorful Stars	SAO 119589, 33 Vir, SAO 119586, SAO 119593, SAO 119606, SAO 119613
Found With	8 x 50 Finderscope
Description	A distinctive binocular asterism with a passing resemblance to the constellation Corona Borealis. The distance from 33 Vir to SAO 119613 is 48'. Just 2 degrees to the west is the "Y" shaped Rho Virginis asterism.

[Star chart showing:
- SAO 119589 (91), 33 Vir A (57), B (91)
- SAO 119613 (77)
- SAO 119582 (78)
- SAO 119606 (89)
- TYC 00882-0482 1 (104)
- SAO 119602 (79)
- SAO 119586 (69)
- TYC 00882-0256 1 (99)
- SAO 119593 (91)]

Principal Stars

SAO 119589 Single Star
aka PPM 159168, BD +10 02471
Magnitude: 9.10

*33 Vir Multiple Star System
aka HR 4849, HD 111028, SAO 119580
Magnitude: 5.65, K0
Distance: 150 +/- 6.5 ly
AB: 5.65+9.1 mag, B=SAO 119578
In 1909: PA 191° Sep 171.50"

SAO 119582 Single Star
aka HD 111043, PPM 159156
Magnitude: 7.81, F0

TYC 00882-0482-1 Single Star
aka SI 76475
Magnitude: 10.41

SAO 119586 Single Star
aka HD 111132, PPM 159164, HIP 62375
Magnitude: 6.89, K0
Distance: 680 +/- 140.0 ly

SAO 119593 Single Star
aka PPM 159173, BD +09 02671
Magnitude: 9.13, K5

SAO 119602 Single Star
aka HD 111309, PPM 159189, HIP 62489
Magnitude: 7.93, F5

SAO 119606 Single Star
aka HD 111369, PPM 159192, HIP 62519
Magnitude: 8.88, K0
Distance: 630 +/- 170.0 ly

SAO 119613 Single Star
aka HD 111498, PPM 159207, HIP 62597
Magnitude: 7.68, K0
Distance: 580 +/- 110.0 ly

Object data from SKYTOOLS™ 2 — used with permission.

The Coronae Virginis Asterism / Virgo / 33 Vir / RA: 12h 46m 23s, Dec: +09° 32' 23"

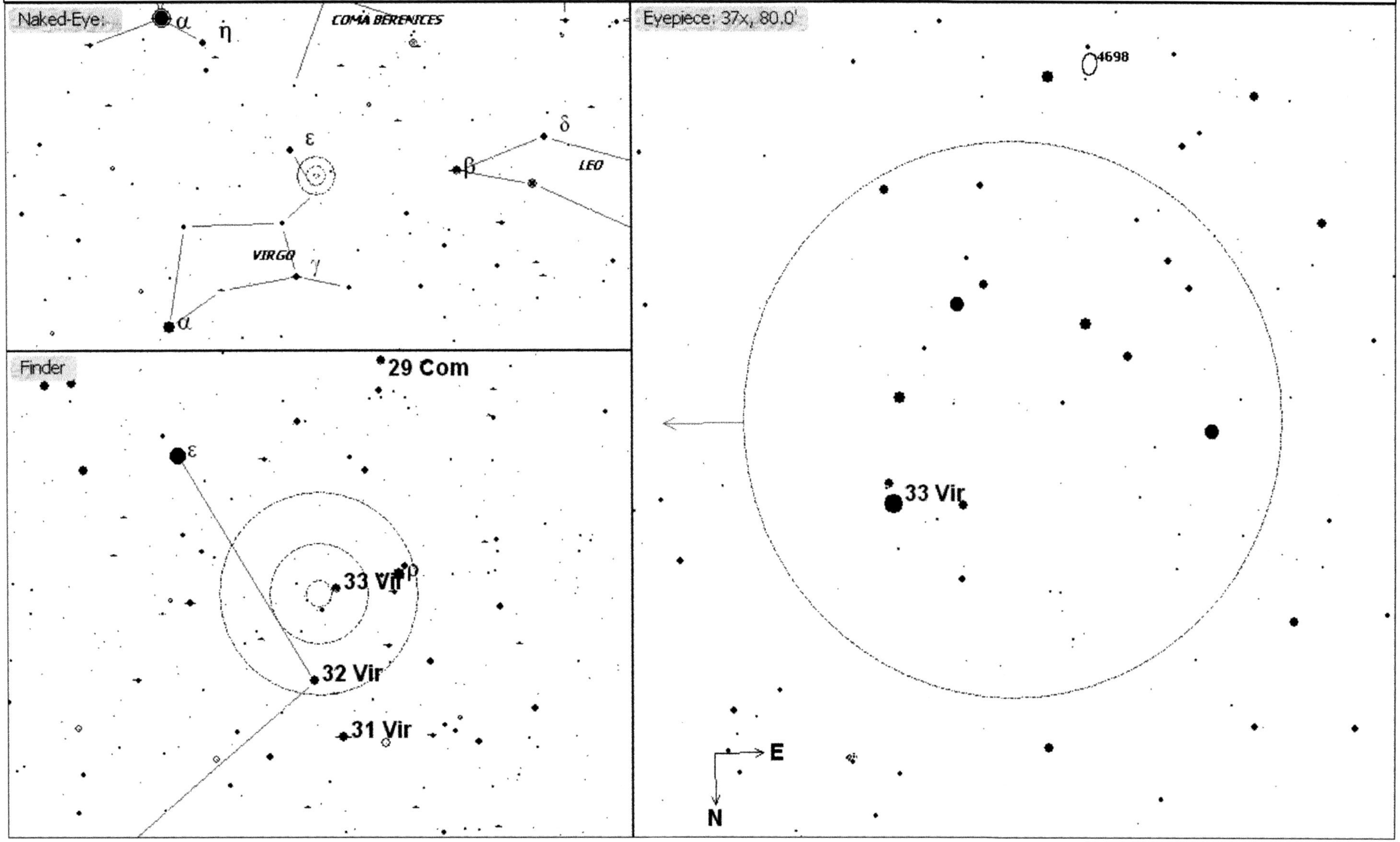

Eyepiece: 37x, 80.0'

Naked-Eye

Finder

6" f/8 Reflector

Name	**The 65-66 Virginis Asterism**
Location	RA: 13h 24m 33s, Dec: -05° 09' 50" (2000) in Virgo
Colorful Stars	SAO 139326, SAO 139314, 65 Vir, SAO 139325, SAO 139316, SAO 139322
Found With	8 x 50 Finderscope
Description	A binocular asterism of two bent lines that diverge at 66 Vir. From SAO 139326 one runs 53' to the NW thru 65 Vir and ends at SAO 139299. The other takes a crooked path northward for a degree before ending at SAO 139315.

Principal Stars

SAO 139326 Single Star
aka HD 116609, PPM 196483, BD -04 03474
Magnitude: 8.38, K2
*66 Vir Single Star
aka HR 5050, HD 116568, SAO 139324
Magnitude: 5.76, F3
Distance: 98 +/- 3.9 ly
SAO 139314 Single Star
aka HD 116461, PPM 196471, HIP 65379
Magnitude: 8.26, K0
65 Vir Single Star
aka HR 5047, HD 116365, SAO 139308
Magnitude: 5.88, K0
Distance: 1200 +/- 360.0 ly
SAO 139299 Multiple Star System
aka HD 116209, PPM 196439, HIP 65243,
BU 1084, ADS 8873
Magnitude: 7.79, F8
AB: 7.79+13.9 mag
In 1889: PA 89° Sep 2.70"

SAO 139325 Single Star
aka HD 116582, PPM 196482, BD -04 03473
Magnitude: 7.40, K2
SAO 139321 Single Star
aka HD 116546, PPM 196478, BD -03 03461
Magnitude: 9.00, G0
SAO 139316 Single Star
aka HD 116501, PPM 196474, HIP 65392
Magnitude: 7.73, K0
Distance: 870 +/- 240.0 ly
SAO 139322 Multiple Star System
aka HD 116545, PPM 196479, HIP 65414
Magnitude: 6.82, K0
Distance: 390 +/- 53.0 ly
AB: 6.82+10.85 mag;
In 1991: PA 239° Sep 0.40"
SAO 139315 Single Star
aka HD 116500, PPM 196473, BD -03 03460
Magnitude: 9.56, A0

Object data from SKYTOOLS™ 2 — used with permission.

The 65 - 66 Virginis Asterism / Virgo / 66 Vir / RA: 13h 24m 33s, Dec: -05° 09' 50"

Eyepiece: 37x, 80.0'

66 Vir
65 Vir

E / N

Naked-Eye

Finder

44 Vir
VIRGO
θ
66 Vir
80 Vir
74 Vir
82 Vir
ζ

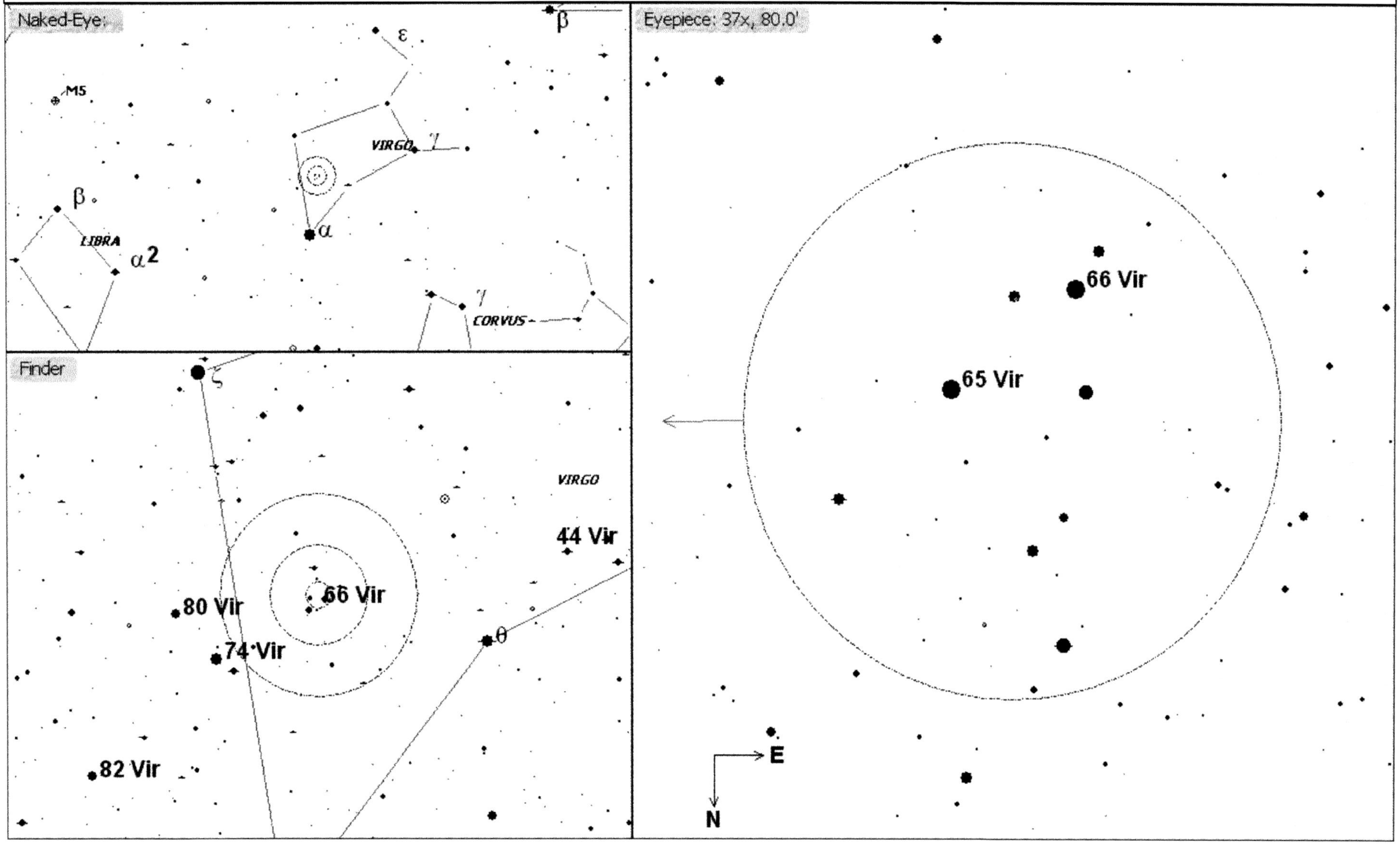

6" f/8 Reflector

Name	**M104 Eastern Asterism**
Location	RA: 12h 43m 48s, Dec: -12° 00 '51" (2000) in Corvus
Colorful Stars	SAO 157473 'A'
Found With	12.5" Dobsonian
Description	A triangular asterism located 1 degree southeast of M104, the Sombrero Galaxy. The distance between SAO 157463 and TYC 05531-0844-1 is just over 15'.

.TYC 05531-0844 1 (98)

.SAO 157473 A (66)   .SAO 157463 (100)

.SAO 157475 (97)
.TYC 05531-0187 1 (107)

.TYC 05531-1360 1 (106)

Principal Stars

SAO 157463 Single Star
aka HD 110559, PPM 226242,
TYC 05531-0205-1
Magnitude: 9.96, G5
TYC 05531-0187-1 Single Star
aka PPM 717087
***SAO 157473** Multiple Star System
aka HD 110662, PPM 226262, HIP 62109,
ADS 8645, HU 738, CCDM 12438-1201
Magnitude: 6.61, K0
Distance: 540 +/- 83.0 ly
AB: 6.61+11.8 mag
In 1959: PA 259° Sep 9.10"

TYC 05531-1360-1 Single Star
aka PPM 717088, SI 581822
Magnitude: 10.55
SAO 157475 Single Star
aka HD 110663, PPM 226265,
TYC 05531-1091-1
Magnitude: 9.66, G0
TYC 05531-0844-1 Single Star
aka PPM 226272 BD -11 03357, SI 581795
Magnitude: 9.76

Object data from SKYTOOLS™ 2 — used with permission.

The M104 Eastern Asterism / Virgo / SAO 157473 / RA: 12h 43m 48s, Dec: -12° 00' 51"

Naked-Eye

Eyepiece: 60x, 50.0'

Finder

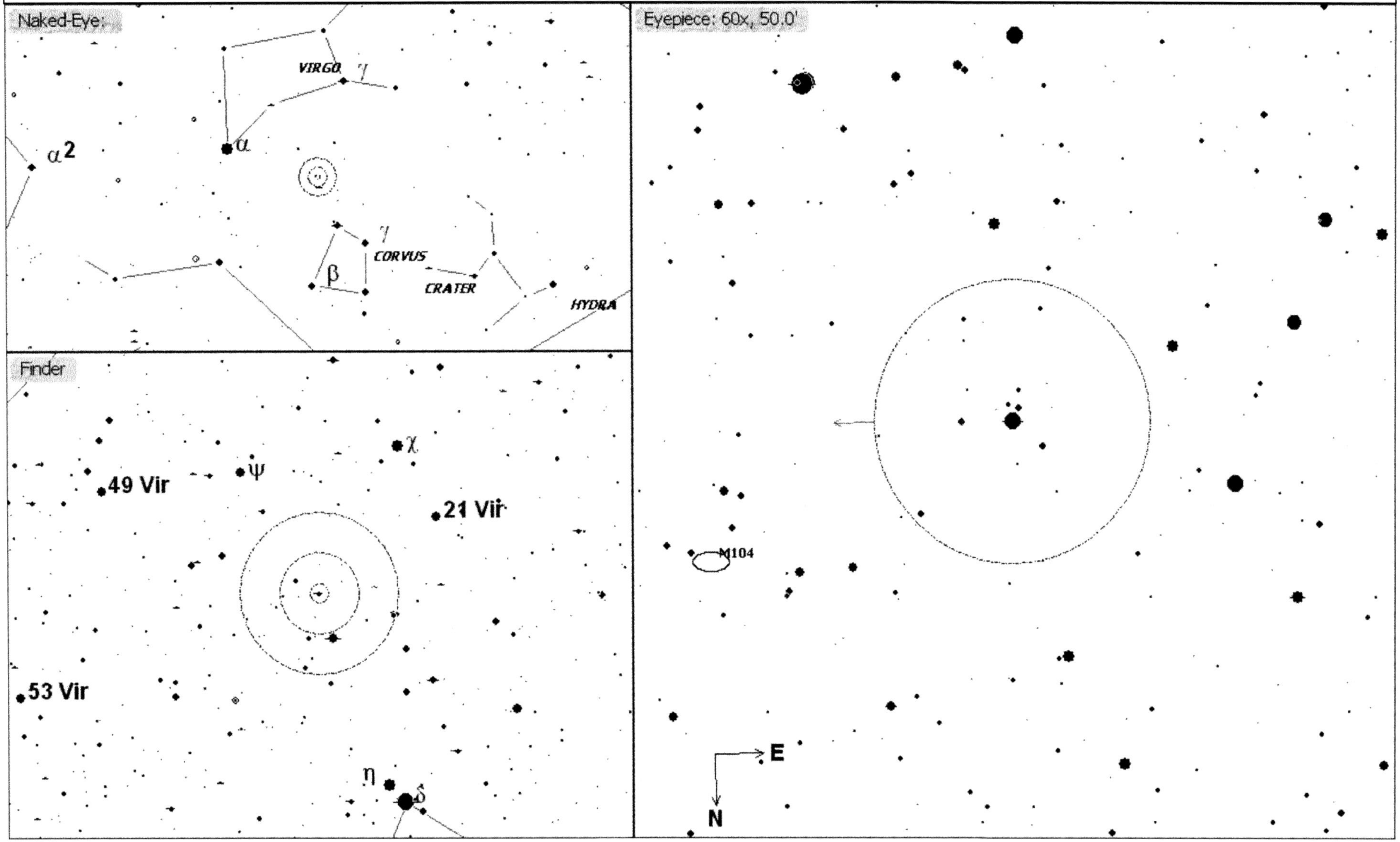

6" f/8 Reflector

Name	**The M 104 Western Asterism / Struve 1664**
Location	RA: 12h 38m 20s, Dec: -11° 31' 01" (2000) in Virgo
Colorful Stars	Σ1664 'A' , SAO 157412
Found With	12.5" Dobsonian
Description	An asterism of six stars about 25' west and slightly north of M104, the Sombrero Galaxy. Σ 1664 'C' and TYC 05531-1156-1 will appear as one star at low power. This asterism is five arc-minutes in length. The wonderful M104 is in the top "10" galaxies of all time.

[Star chart showing asterism with labeled stars: TYC 05531-1156 1 (111), C (115), D (116), SAO 157411 A (77), B (93), SAO 157412 (85), SAO 157413 (89)]

Principal Stars

TYC 05531-1156 1 Single Star
aka SI 581811
Magnitude: 11.11

* Σ 1664 Multiple Star System
aka Struve 1664, SAO 157411, HD 109875, PPM 226124
Magnitude: 7.66, K0 (A Component)
Distance: 1300 +/- 570.0 ly

SAO 157412 Single Star
aka HD 109899, PPM 226128
Magnitude: 8.47, G5

SAO 157413 Single Star
aka HD 109916, PPM 226132, BD -10 03515
TYC 05531-1324 1
Magnitude: 8.88, A2

Sombrero Galaxy
aka M 104, NGC 4594, PGC 42407
Magnitude: 9.10
Size: 8.3'x 4.1'
Hubble Class: Spiral ; Orientation: Edge on
Position Angle: 89°

Σ 1664 Components

AB:	7.7+9.3 mag	In 1923: PA 237° Sep 26"	B= HD 109874
AC	7.7+11.5 mag	In 1907: PA 306° Sep 62"	
CD	11.5+11.6 mag	In 1907: PA 266° Sep 32"	

All information from SKYTOOLS™ 2 — used with permission.

The M104 Western Asterism / Virgo / Σ1664 / RA: 12h 38m 20s, Dec: -11° 31' 01"

Naked-Eye

Finder

Eyepiece: 60x, 50.0'

6" f/8 Reflector

Name	**The Syrma Asterism**
Location	RA: 14h 16m 01s, Dec: -06° 00' 02" (2000) in Virgo
Colorful Stars	SAO 139840, 139853
Found With	8 x 50 Finderscope
Description	A fuzzy spot in the finder, the stars around Iota Virginis form a triangular asterism. Each corner is a pair of stars. A line of stars runs along the northeast side of the triangle, which along with the pair north of CF Vir, gives the impression of an open cluster at low power.

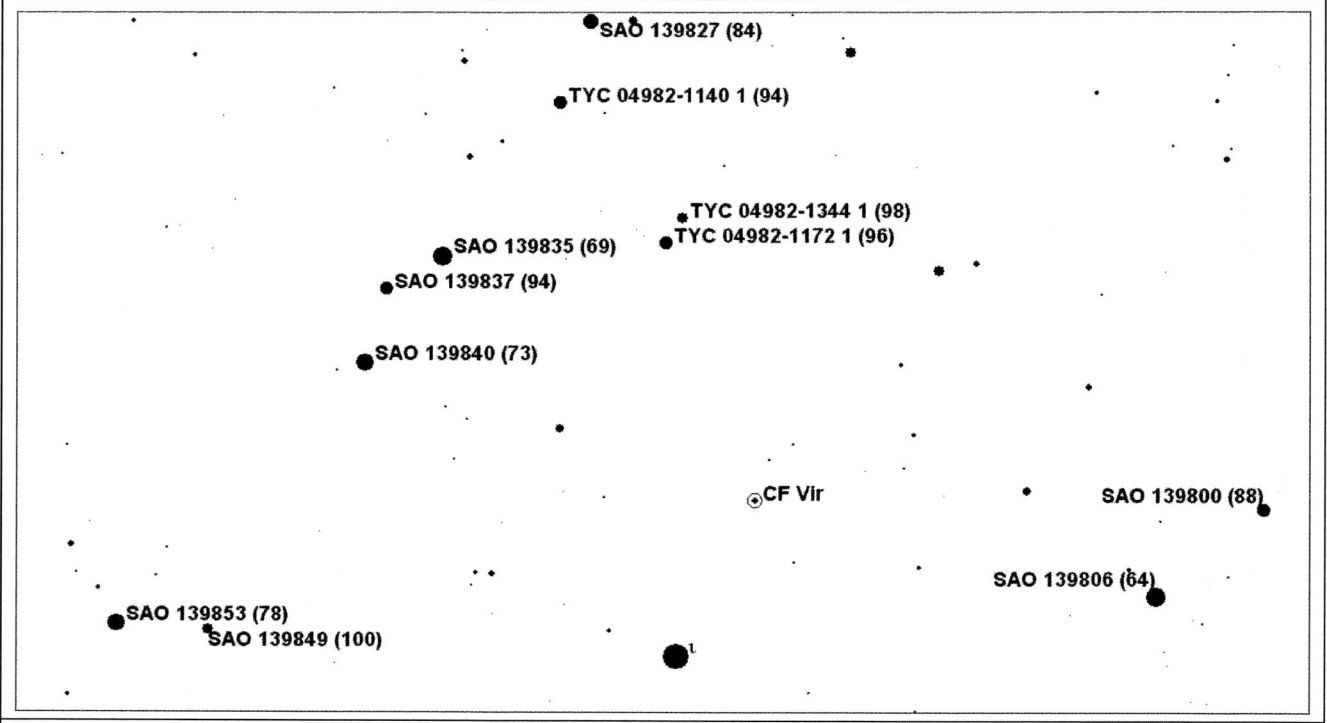

Principal Stars

SAO 139806 Single Star
aka HR 5322 HD 124553, PPM 197336
Magnitude: 6.36, F9
Distance: 140 +/- 6.0 ly
CF Vir Variable Star
aka GCVS 28028, GSC 04982-1186
Variability: Mira (o) Ceti-type variable
Mag: P11.00 to P14.00
Period: 227.600006 days
Epoch: JD 2431970.0000
*Syrma Variable Star
aka Iota Vir, 99 Vir, HR 5338, HD 124850
Magnitude: V 4.04 to V 4.11, F7
Distance: 70 +/- 1.3 ly
TYC 04982-1344-1 Single Star
aka HD 124836, PPM 706012,
Magnitude: 9.83
TYC 04982-1172-1 Single Star
aka HD 124866, PPM 197359
Magnitude: 9.56

SAO 139827 Single Star
aka HD 124887, PPM 197363
Magnitude: 8.41, A3
TYC 04982-1140-1 Single Star
aka HD 124901 PPM 197369
Magnitude: 9.37
SAO 139835 Single Star
aka HD 124988, PPM 197375, HIP 69775
Magnitude: 6.88, F0
SAO 139837 Single Star
aka PPM 197377, BD -05 03847
Magnitude: 9.43, G0
SAO 139840 Single Star
aka HD 125043, PPM 197380, HIP 69798
Magnitude: 7.34, K0
SAO 139853 Single Star
aka HD 125183, PPM 197404, HIP 69875
Magnitude: 7.77, K2
Distance: 1300 +/- 560.0 ly
SAO 139849 Single Star
aka HD 125124, PPM 197395
Magnitude: 9.96, F8

Object data from SKYTOOLS™ 2 — used with permission.

The Syrma Asterism / Virgo / Iota Vir / RA: 14h 16m 01s, Dec: -06° 00' 02"

Eyepiece: 37x, 80.0'

Naked-Eye

Finder

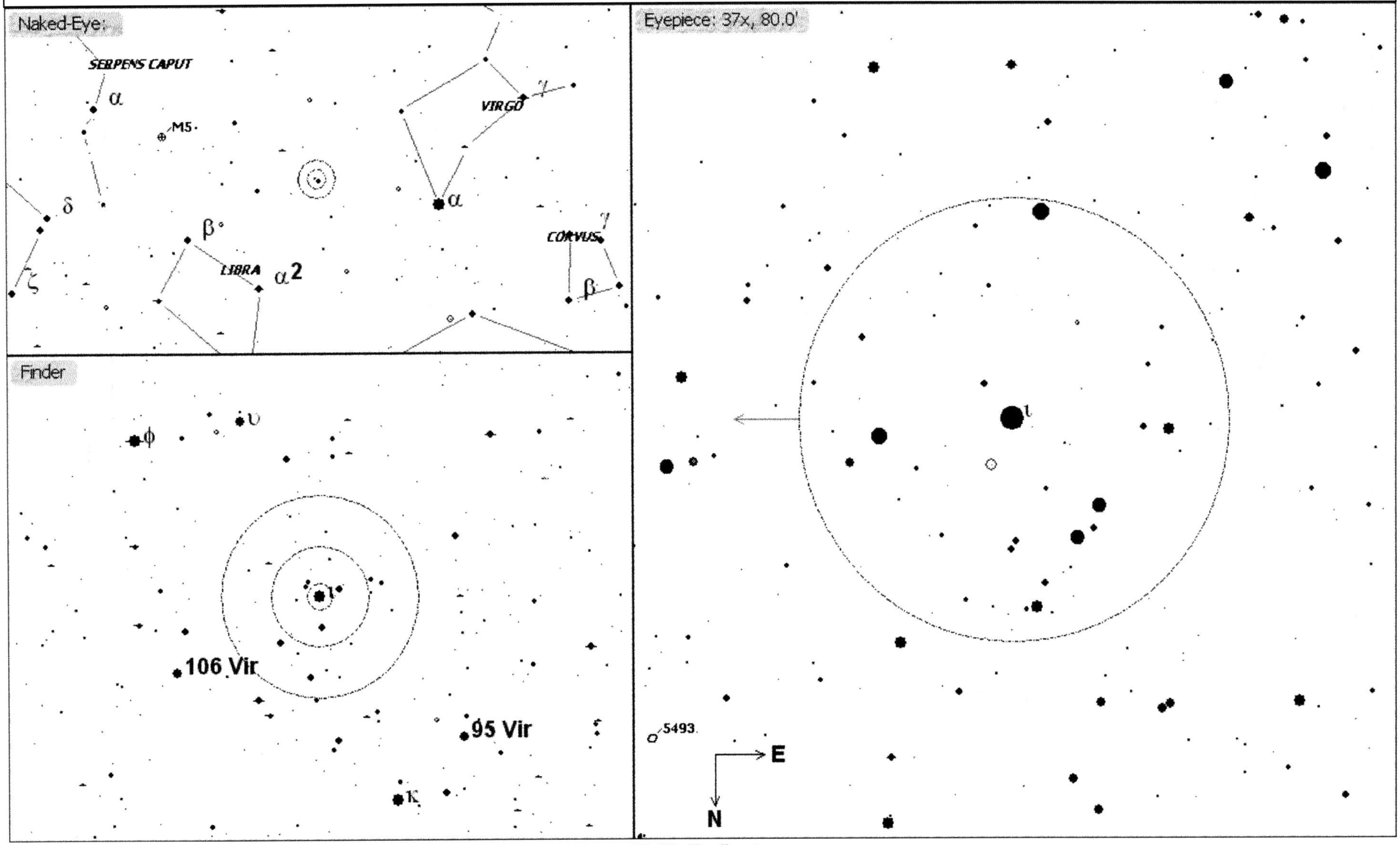

6" f/8 Reflector

173

Name	**The Coathanger / Collinder 399**
Location	RA: 19h 26m 13.2s, Dec: +20° 05' 52" (2000) in Vulpecula
Colorful Stars	SAO 87140, 104839, 4 Vul
Found With	Direct Vision
Description	The Coathanger is a bright visual fuzzy object in the summer Milky Way. Unlike the Beehive or the Coma Cluster, it's not a real open cluster. The Coathanger is always a favorite at star parties. The faint and real open cluster NGC 6802 is easily found just east of 7 Vul. The distance from SAO 87140 to 7 Vul is 1.5 degrees.

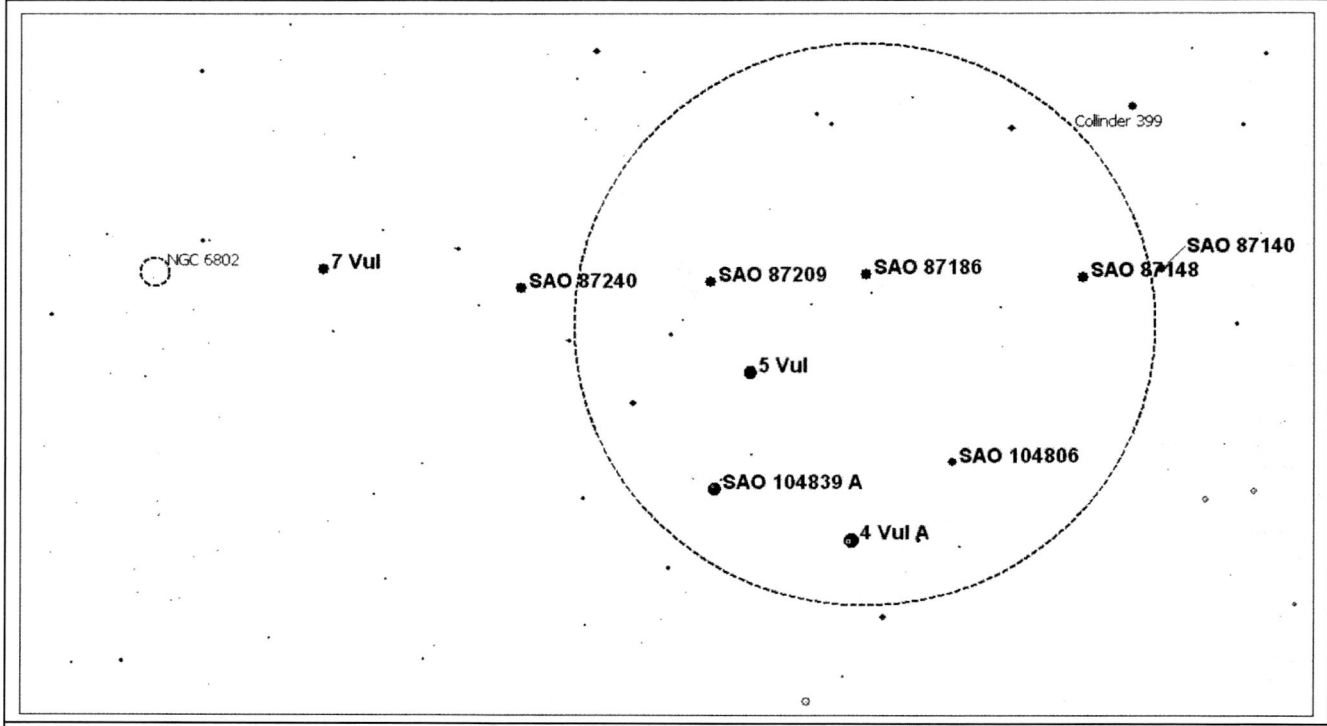

Principal Stars

Coathanger Bar	Coathanger Hook
SAO 87140 Single Star aka HD 182293, PPM 108893, HIP 95291 Magnitude: 7.11, K1 Distance: 360 +/- 31.0 ly	***5 Vul** Single Star aka HR 7390, HD 182919, SAO 104831 Magnitude: 5.60, B9 Distance: 220 +/- 10.0 ly
SAO 87148 Single Star aka HD 182422, HR 7364, PPM 108912 Magnitude: 6.40, B8 Distance: 1100 +/- 270.0 ly	**SAO 104839** Multiple Star System aka HD 182955, HR 7391, PPM 136291, HIP 95582, Struve 2521, ADS 12445 Variability: Possibly micro-variable. Magnitude: 5.87, M0III Distance: 450 +/- 48.0 ly AB: 5.84+10.8 mag; In 1958: PA 35° Sep 26.70" AC: 5.84+10 mag; In 1908: PA 323° Sep 70.40" AD: 5.84+10 mag, In 1878: PA 64° Sep 149.60"
SAO 87186 Single Star aka HD 182761, HR 7384, PPM 108971 Magnitude: 6.31, A0 Distance: 400 +/- 36.0 ly	
SAO 87209 Single Star aka HD 182972, PPM 109008, HIP 95584 Magnitude: 6.64, A1 Distance: 560 +/- 77.0 ly	**4 Vul** Multiple Star System aka HR 7385, HD 182762, SAO 104818, HJ 2871, ADS 12425 Magnitude: 5.14, K0, Distance: 240 +/- 13.0 ly AB: 5.14+10 mag In 1957: PA 100° Sep 18.90" AC: 5.14+11.7 mag In 1959: PA 204° Sep 52.60"
SAO 87240 Single Star aka HD 183261, PPM 109055, HIP 95700 Magnitude: 6.84, B3 Distance: 900 +/- 200.0 ly	
7 Vul Single Star aka HR 7409, HD 183537, SAO 87269 Variability: Unsolved variable. Magnitude: 6.34, B5 Distance: 760 +/- 130.0 ly	**SAO 104806** Single Star aka HD 182620, PPM 136248, HIP 95432 Magnitude: 7.16, A2 Distance: 550 +/- 76.0 ly
NGC 6802 Open Cluster aka Collinder 400, OCL 114 Magnitude: 11.70, Size: 3.2' Distance: 3200 ly Age: 1080 Myrs	**Collinder 399** Open Cluster aka OCL 113 Magnitude: 4.80, Size: 60.0' Distance: 420 ly, Age: 200 Myrs

Object data from SKYTOOLS™ 2 — used with permission.

The Coathanger / Collinder 399 / Vulpecula / 5 Vul / RA: 19h 26m 13s, Dec: +20° 05' 52"

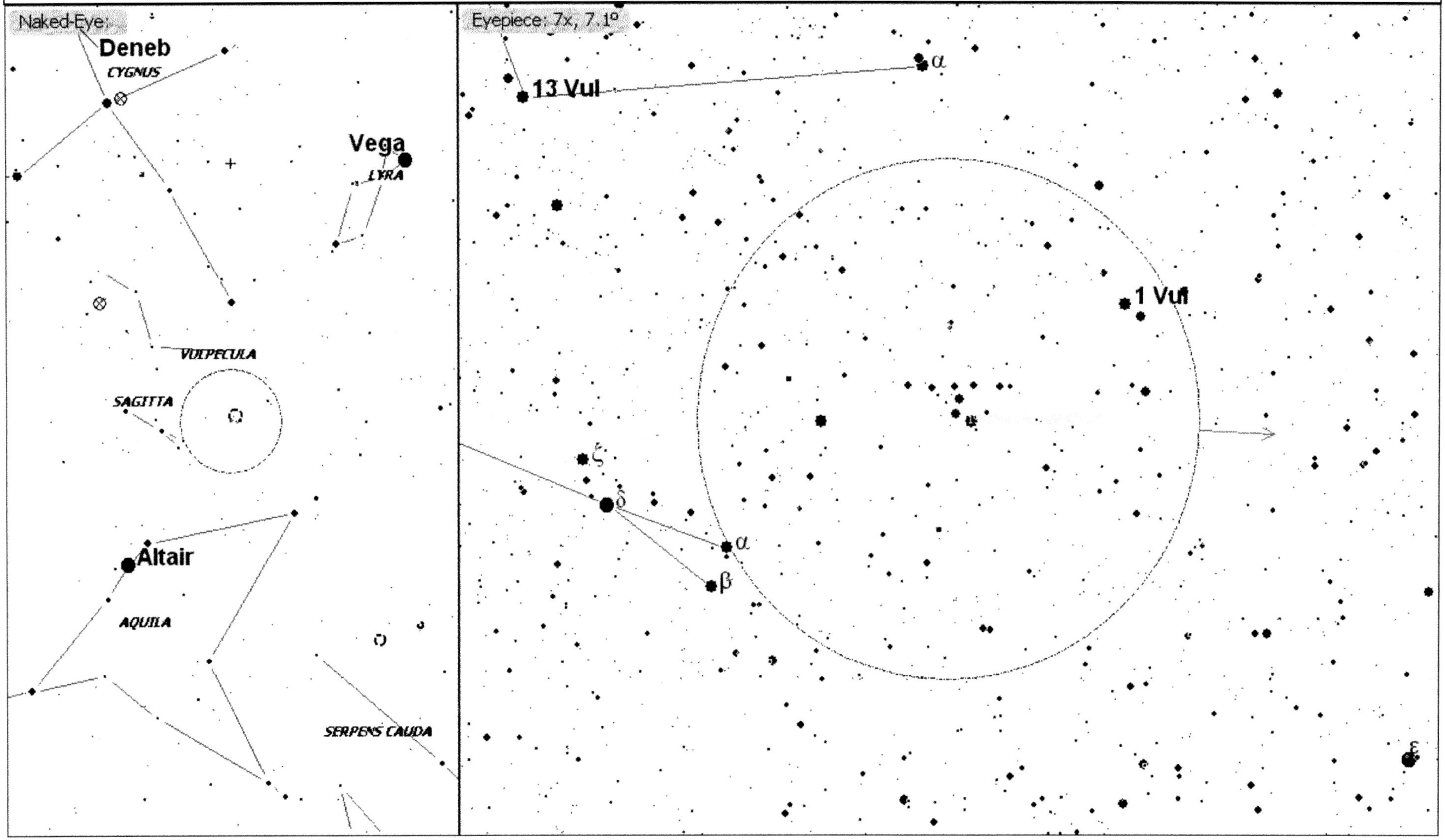

Eyepiece: 7x, 7.1°

1 Vul
α
13 Vul
α
β
ζ
δ
ε

Naked-Eye:

Deneb
CYGNUS
Vega
LYRA
VULPECULA
SAGITTA
Altair
AQUILA
SERPENS CAUDA

7 x 50 Binocular

Notes

Notes

Notes